Formulations of Classical and Quantum Dynamical Theory

This is Volume 60 in
MATHEMATICS IN SCIENCE AND ENGINEERING

A series of monographs and textbooks
Edited by RICHARD BELLMAN, *University of Southern California*

A complete list of the books in this series appears at the end of this volume.

Formulations of Classical and Quantum Dynamical Theory

Gerald Rosen

Department of Physics
Drexel Institute of Technology
Philadelphia, Pennsylvania

 ACADEMIC PRESS   *New York and London*   *1969*

ACADEMIC PRESS, INC.
111 Fifth Avenue, New York, New York 10003

United Kingdom Edition published by
ACADEMIC PRESS, INC. (LONDON) LTD.
Berkeley Square House, London W1X 6BA

LIBRARY OF CONGRESS CATALOG CARD NUMBER: 76–97481

PRINTED IN THE UNITED STATES OF AMERICA

To Sarah

Contents

Preface   xi

Introduction   1

1. **Classical Mechanics**

    1. Lagrange Formulation   4
    2. Hamilton Formulation   8
    3. Poisson Formulation   12

2. **Quantum Mechanics**

    1. Feynman Formulation   22
    2. Schrödinger Formulation   34
    3. Dirac Formulation   41

3. **Classical Field Theory**

    1. Neologized Lagrange Formulation   52
    2. Neologized Hamilton Formulation   57
    3. Neologized Poisson Formulation   60

4. **Quantum Field Theory**

1. Neologized Feynman Formulation   68
2. Neologized Schrödinger Formulation   77
3. Neologized Dirac Formulation   85

Prospects   93

APPENDICES

A   **Functional Differentiation**   96

B   **Linear Representations of a Lie Group**   101

C   **Haar Measure**   110

D   **Functional Integration by Parts Lemma**   116

E   **Relativistic Sum-Over-Histories for the
    One-Dimensional Dirac Equation**   118

F   **Feynman Operators**   123

G   **Quantum and Classical Statistical Mechanics**   127

H   **The Iteration Solution of Perturbation Theory**   131

I   **Existence of Spatially Localized Singularity-Free
    Periodic Solutions in Classical Field Theories**   134

J   **Rigorous Solutions in Essentially Nonlinear
    Classical Field Theories**   137

Contents

K   **Quantum Theory of Electromagnetic Radiation**   140

L   **Stationary States in Quantum Field Theories**   145

Index   149

Preface

For the past several years, I have given a course to seniors and graduate students on the mathematical structure within and the logical relationships between classical mechanics and quantum mechanics for nondissipative closed physical systems. This monograph has developed from my lecture notes for that course.

Classical mechanics is formulated according to Lagrange, Hamilton, and Poisson; quantum mechanics, correspondingly, is formulated according to Feynman, Schrödinger, and Dirac. Because of personal predilection and the feeling that something is missing in previous treatments of the subject, I emphasize and discuss in detail the Feynman passage and "sum-over-histories" formulation of quantum mechanics. The essential mathematics, the calculus of functional differentiation and integration and the theory of Lie groups and algebras, is introduced in elementary terms in order to make this a self-contained exposition and one that is readily accessible to readers familiar with the concepts of ordinary calculus and differential equations. The appendices include brief introductions to more advanced mathematics intrinsic to this subject, along with a number of applications of functional calculus methods for quantum fields.

No attempt has been made to present an exhaustive account of the mathematical framework in classical and quantum dynamical theory, nor to augment what we do discuss with complete rigor in the sense of modern mathematics. Moreover, in order to render a unified quality to the work, I have omitted topics that might have been included, such as Hilbert space and general representation theory in quantum mechanics. My hope is that the references cited will be adequate to guide the reader to more specialized works. However, the references that appear here are merely representative of the very extensive postwar literature pertinent to the foundations of classical and quantum dynamical theory.

It is a pleasure to thank Professor Richard Bellman for his kind encouragement to compose this monograph. I also wish to acknowledge the valuable contribution made by students who have critically examined the subject matter in my course.

GERALD ROSEN

Philadelphia, Pennsylvania
September, 1969

Formulations of
Classical and Quantum
Dynamical Theory

Introduction

Nature exhibits a picture of myriad *form* and continual *change*. In a broad sense, the object of scientific theory has been twofold: (1) to determine the constitution or composition of observable things, and (2) to describe and predict how observable things move or change with time.

There is a strong interdependency between the theory of form and the theory of change manifest in astronomy, biology, chemistry, and all other branches of science, although varying emphasis may be given to the former or the latter. The description of time evolution is accessible to precise mathematical formulation in the domain of physics, where phenomena can be understood in terms of the constitution and time evolution of the most elementary observable things. Since all observable things in nature are built up from the elementary observable things studied in physics, the description and prediction of time evolution in nature reduces in principle to dynamical theory in physics.

Fundamental dynamical theory in physics is formulated on two distinct levels: (1) classical mechanics primarily for large-scale phenomena and (2) quantum mechanics primarily for small-scale (atomic) phenomena.[1] The inception of classical mechanics came in 1687 with Newton's *Principia*, but it was not until the eve of the French Revolution that the mathematical description achieved full flower with Lagrange's *Méchanique Analytique* in 1788. Subsequently, alternative formulations

---

[1] Size-scale of the phenomena is not a completely accurate index because we have macroscopic quantum theories for dense boson fluids and for superconductivity.

1

of Lagrange's mathematical theory by Hamilton and Poisson appeared during the period 1810–1819. In contrast to classical mechanics, the mathematical foundations of quantum mechanics were already evident during 1925–1929 in the papers of Schrödinger, Dirac, Heisenberg, and Pauli, although two decades elapsed before the important alternative Feynman formulation was communicated widely in 1948.

This monograph gives a self-contained and logically unified exposition of conceptual framework in classical mechanics and quantum mechanics, for particles and for fields. The physical systems to which the formalism applies are *nondissipative* (that is, a record in time of any observed motion admits a conjugation transformation such that the conjugated record is admissible dynamically if read backward in time) and *closed* (that is, no time-dependent external forces act on the system and influence the dynamics). For such nondissipative, closed physical systems with a classical analog, the main relationships that appear between the two fundamental dynamical theories are emphasized, and a unified account of the ways to make passage from the classical to the quantum theory is given, along with how the different ways are interrelated logically. Summarizing the content of Chapters 1 and 2,[2] the outline in the schematic diagram below is covered for dynamical systems with a finite number of degrees of freedom, as exemplified by systems composed of a finite number of interacting particles:

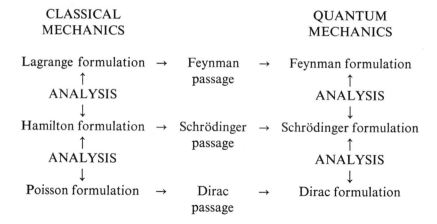

CLASSICAL MECHANICS		QUANTUM MECHANICS
Lagrange formulation →	Feynman passage →	Feynman formulation
↑		↑
ANALYSIS		ANALYSIS
↓		↓
Hamilton formulation →	Schrödinger passage →	Schrödinger formulation
↑		↑
ANALYSIS		ANALYSIS
↓		↓
Poisson formulation →	Dirac passage →	Dirac formulation

[2] An abbreviated presentation of Chapters 1 and 2 appeared in the June 1968 issue of the *J. Franklin Inst.* **285**, 409–423.

A detailed and up-to-date discussion of the conceptually paramount Feynman passage and " sum-over-histories" formulation for quantum mechanics appears in Chapter 2. Quantum dynamical formulations are also given for Pauli's neutrino and Dirac's electron, important dynamical systems which have no analogs in classical mechanics. The development in Chapters 1 and 2 is paraphased for fields, continuous systems with an infinite number of degrees of freedom, in Chapters 3 and 4. Because we work with the "coordinate-diagonal" representation for quantum fields with a classical analog in Chapter 4, a simple mathematical hierarchy in the character of equations appears, which in the differential dynamical laws for the systems is displayed by the schematic diagram:

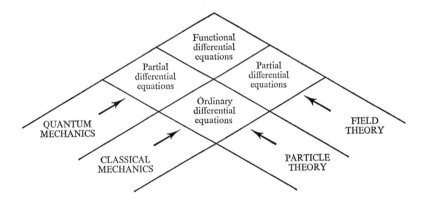

Mathematical topics which might be postponed at a first reading have been relegated to the appendices. In the latter appendices, we present some specialized applications of the formalism for quantum field theory. Here, the mathematical *leit motiv* throughout our exposition of fundamental dynamical theory, the calculus of functional differentiation and integration, emerges as the essential and primary vehicle for analysis.

Chapter 1 *Classical Mechanics*

1. Lagrange Formulation

Taken as a primitive physical notion, the *state* of a dynamical system with a finite number of degrees of freedom is realized at any instant of time by fixing the components of $q = (q_1, \ldots, q_n)$, the *generalized coordinate* real *n-tuple*, and $\dot{q} = (\dot{q}_1, \ldots, \dot{q}_n)$, the *generalized velocity* real *n-tuple*. An "*n*-tuple," synonymous with "$1 \times n$ matrix," is simply an ordered array of n (real or complex) numbers; the generic "*n*-tuple" is to be distinguished from the generic "*n*-dimensional vector," the latter term, synonymous with "rank one *n*-tensor," reserved for mathematical objects which have a certain linear transformation character in physical theories. No special transformation character need be featured by the generalized coordinate and generalized velocity *n*-tuples, and the components of q (or $\dot{q}$) may exhibit different physical dimensions, as in the case of curvilinear coordinates. A q with each component expressible as a linear combination of Cartesian coordinates is called a *linear coordinate n-tuple*. Thus, for a dynamical system composed of r structureless (point-mass) particles with masses $m_1, \ldots, m_r$ and Cartesian coordinates $x_1, y_1, z_1, \ldots, x_r, y_r, z_r$, it is usually convenient to work with the linear coordinate $3r$-tuple

$$q = [(m_1)^{1/2}x_1, (m_1)^{1/2}y_1, (m_1)^{1/2}z_1, \ldots, (m_r)^{1/2}x_r, (m_r)^{1/2}y_r, (m_r)^{1/2}z_r],$$

4

so that the kinetic energy of the dynamical system assumes the simple form $\frac{1}{2}\dot{q} \cdot \dot{q} \equiv \frac{1}{2}\sum_{i=1}^{n}(\dot{q}_i)^2$; for a dynamical system composed of two structureless particles constrained to move along the x-axis, it is often most convenient to analyze the dynamics in terms of the linear coordinate 2-tuple $q = [(m_1 + m_2)^{1/2}(x_1 + x_2), x_1 - x_2]$.

The system *dynamics*, again a primitive physical notion, is described by a function of time, $q = q(t)$ (function class C^1, piecewise C^2). This gives the evolution of the generalized coordinate n-tuple; the generalized velocity n-tuple is then required to take the associated form $\dot{q} = dq(t)/dt$. Finally, physically admissible functions $q = q(t)$ are prescribed by a *dynamical law* in the form of *Euler–Lagrange equations*

$$\partial L/\partial q - (d/dt)(\partial L/\partial \dot{q}) = 0, \tag{1.1}$$

where the *Lagrangian* $L = L(q, \dot{q})$ is a real scalar (1-tuple) function of q and $\dot{q}$; in (1.1), n-tuple gradient operators appear as

$$\partial/\partial q = (\partial/\partial q_1, \ldots, \partial/\partial q_n) \quad \text{and} \quad \partial/\partial \dot{q} = (\partial/\partial \dot{q}_1, \ldots, \partial/\partial \dot{q}_n).$$

Consider, for example, the Lagrangian

$$L = \frac{1}{2}\dot{q} \cdot \dot{q} - V \equiv \frac{1}{2}\sum_{i=1}^{n}(\dot{q}_i)^2 - V, \tag{1.2}$$

where a dot between n-tuples denotes contraction and the *potential energy* $V = V(q)$ is a real 1-tuple function of q; putting (1.2) into (1.1) produces the Newtonian form of the Euler–Lagrange equations,

$$\ddot{q} + \partial V/\partial q = 0. \tag{1.3}$$

Since the Euler–Lagrange equations, and not the Lagrangian, are directly associated with observed motion, we must consider whether a given set of Euler–Lagrange equations such as (1.3) requires a specific and essentially unique Lagrangian.[1] We have the obvious nonuniqueness of the Lagrangian stemming from Eq. (1.1) being invariant with respect to transformations of the form

$$L \to (L + \partial\chi/\partial q \cdot \dot{q}) \equiv \left(L + \sum_{i=1}^{n}(\partial\chi/\partial q_i)\dot{q}_i\right), \tag{1.4}$$

---

[1] We are concerned here with the general invariance of Eqs. (1.1) and not the covariance associated with the canonical transformations discussed later in this chapter. For example, a covariance generated by a transformation of the form $L(q, \dot{q}) \to L(\Lambda q, \Lambda\dot{q})$ with Λ an arbitrary nonsingular constant $n \times n$ matrix (and $(\Lambda q)_i \equiv \sum_{j=1}^{n} \Lambda_{ij}q_j$) is of no interest here.

where $\chi = \chi(q)$ is an arbitrary function of q, but this nonuniqueness[2] has trivial physical consequences and can be waived mathematically by simply evoking the definition that Lagrangians related by a transformation (1.4) are "dynamically equivalent." Moreover, the *canonical energy* integral to (1.1),

$$E \equiv \dot{q} \cdot \partial L / \partial \dot{q} - L, \tag{1.5}$$

a quantity that is a constant of the motion, $\dot{E} = 0$ for $q = q(t)$ satisfying (1.1), is identical for all Lagrangians which are related by a transformation (1.4). Are there *inequivalent Lagrangians* [unrelated by transformations of the form (1.4)] that produce the same prescribed set of Euler–Lagrange equations? To avoid undue technical complications, let us consider the Newtonian Euler–Lagrange equations (1.3). Necessary conditions for (1.3) to be compatible with (and follow from) (1.1) are obtained as a system of linear homogenous partial differential equations for $L = L(q, \dot{q})$ by using (1.3) to eliminate $\ddot{q}$ from (1.1),

$$\left\{ \frac{\partial}{\partial q_i} + \left(\frac{\partial V}{\partial q} \cdot \frac{\partial}{\partial \dot{q}} - \dot{q} \cdot \frac{\partial}{\partial q} \right) \frac{\partial}{\partial \dot{q}_i} \right\} L = 0. \tag{1.6}$$

In the general case with $n > 1$, we have more equations in the system (1.6) than unknown 1-tuple real functions, namely, L. Moreover, the integrability conditions for the system (1.6), obtained by evaluating commutators of the curly-bracketed differential operators,

$$\sum_{k=1}^{n} \left(\frac{\partial^2 V}{\partial q_i \, \partial q_k} \frac{\partial^2 L}{\partial \dot{q}_k \, \partial \dot{q}_j} - \frac{\partial^2 V}{\partial q_j \, \partial q_k} \frac{\partial^2 L}{\partial \dot{q}_k \, \partial \dot{q}_i} \right)$$
$$+ \left(\frac{\partial V}{\partial q} \cdot \frac{\partial}{\partial \dot{q}} - \dot{q} \cdot \frac{\partial}{\partial q} \right) \left(\frac{\partial^2 L}{\partial \dot{q}_i \, \partial q_j} - \frac{\partial^2 L}{\partial \dot{q}_j \, \partial q_i} \right) = 0, \tag{1.7}$$

are not satisfied identically or as a consequence of (1.6) for general $V(q)$. In fact, the combined set of Eqs. (1.6) and (1.7) is still not a complete system, additional linearly independent equations for L being obtained by evaluating commutators of the differential operators in (1.6) and (1.7). Hence, we have more than $\frac{1}{2}(n^2 + n)$ partial differential equations that must be satisfied simultaneously by L in the general case with $n > 1$. The only known solution to this system with $n > 1$ and general

---

[2] This is the most general modification of L by *an additive function* that leaves the Euler–Lagrange equations invariant; for the proof, see R. Courant and D. Hilbert, "Methods of Mathematical Physics I," p. 193. Wiley (Interscience), New York, 1953.

$V(q)$ is given by (1.2), more precisely, (1.2) multiplied by a constant and modified by a transformation of the form (1.4). However, the case $n = 1$ is exceptional, and for it (1.6) produces the single equation

$$\partial \hat{L}/\partial V + \partial^2 \hat{L}/\partial \dot{q}^2 - \dot{q}(\partial^2 \hat{L}/\partial V \, \partial \dot{q}) = 0, \qquad (1.8)$$

where we have put $L \equiv L(q, \dot{q}) = \hat{L} \equiv \hat{L}(V, \dot{q})$ and divided out the common factor dV/dq that appears in each term. Two arbitrary functions, one of which related to the arbitrary χ in (1.4), are involved in the general integral to the second-order linear homogeneous partial differential equation (1.8),

$$\hat{L} = \xi(\partial/\partial V)[V^{-1} + (2)^{-(1/2)}\dot{q}V^{-(3/2)}\tan^{-1}(\dot{q}/(2V)^{1/2})] + \eta(V)\dot{q}, \quad (1.9)$$

where $\xi(\partial/\partial V)$ is an arbitrary function of the differential operator $\partial/\partial V$ and $\eta(V)$ is an arbitrary function of V. Illustrative special solutions to (1.8), obtained by choosing simplifying forms for ξ in (1.9) and setting $\eta = 0$, are:

$$L = \hat{L} = \tfrac{1}{12}\dot{q}^4 + \dot{q}^2 V - V^2, \qquad (1.10)$$

and

$$L = \hat{L} = -V^{-1}(\tfrac{1}{2}\dot{q}^2 + V)^{1/2}, \qquad (1.11)$$

in addition to the Lagrangian (1.2) with $n = 1$. It is easy to verify that Newton's equation (1.3) follows from the Euler–Lagrange equation (1.1) with $n = 1$ for either (1.10) or (1.11). Although such unfamiliar Lagrangians cannot be excluded on the basis of observed dynamical behavior alone, the canonical energy integrals (1.5) for either (1.10) or (1.11) are not quantities that are likely to equal the physical energy that is attributed to the system. By identifying the physical energy as $(\tfrac{1}{2}\dot{q}^2 + V)$ and postulating that it should equal the canonical energy (1.5), the familiar Lagrangian (1.2) is singled out from the manifold of possible Lagrangians that satisfy Eq. (1.8) and produce Eq. (1.3) with $n = 1$.

Let us now introduce the *action functional* defined by

$$S = S[q(t)] \equiv \int_{t'}^{t''} L(q, \dot{q}) \, dt \qquad (1.12)$$

on the domain of all real n-tuple functions $q(t)$ (of continuity class C^1) that reduce to the prescribed boundary values $q(t') = q'$, a fixed n-tuple, and $q(t'') = q''$, a fixed n-tuple. The domain of S is thus a *class* of real n-tuples functions and not a *space* because the sum of two n-tuple functions $q(t)$ in the domain of S, in general, is not in the domain of S

(that is, the domain of S is not closed under n-tuple addition). The difference of any two $q(t)$ in the domain of S is a real n-tuple in the function space $\mathscr{F} = \{\sigma(t): \sigma(t) \in C^1 \text{ for } t' \leqslant t \leqslant t'', \sigma(t') = \sigma(t'') = 0\}$. For any $\sigma(t) \in \mathscr{F}$, we have

$$\left(\frac{d}{d\varepsilon} S[q(t) + \varepsilon\sigma(t)]\right)_{\varepsilon=0} \equiv \left(\frac{d}{d\varepsilon} \int_{t'}^{t''} L(q + \varepsilon\sigma, \dot{q} + \varepsilon\dot{\sigma}) \, dt\right)_{\varepsilon=0}$$

$$= \int_{t'}^{t''} \left(\sigma \cdot \frac{\partial L}{\partial q} + \dot{\sigma} \cdot \frac{\partial L}{\partial \dot{q}}\right) dt$$

$$= \int_{t'}^{t''} \sigma(t) \cdot \left(\frac{\partial L}{\partial q} - \frac{d}{dt}\frac{\partial L}{\partial \dot{q}}\right) dt, \qquad (1.13)$$

while the functional derivatives of S with respect to the components of q follow from the implicit definition given in Appendix A as

$$((d/d\varepsilon) S[q(t) + \varepsilon\sigma(t)])_{\varepsilon=0} \equiv \int \sigma(t) \cdot [\delta S/\delta q(t)] \, dt. \qquad (1.14)$$

Hence, by comparing (1.13) and (1.14), we find that the functional derivatives of S with respect to the components of $q(t)$ are given explicitly as

$$\delta S/\delta q(t) = \partial L/\partial q - (d/dt)(\partial L/\partial \dot{q}). \qquad (1.15)$$

Therefore, the n second-order Euler–Lagrange equations (1.1) are equivalent to the *action principle*: S *has an extremum at a physical* $q(t)$,

$$\delta S/\delta q(t) = 0, \qquad (1.16)$$

with the left-hand side evaluated at a physical $q(t)$. The action principle (1.16) constitutes the most compact and fundamental statement of the dynamical law.

2. Hamilton Formulation

Although the action principle (1.16) is the most fundamental statement of the dynamical law, there is an important subsidiary formulation which follows directly from (1.1). First, the *generalized momentum* real n-tuple

$$p \equiv \partial L/\partial \dot{q} \qquad (1.17)$$

is introduced, and the *Hamiltonian*

$$H = H(q, p) \equiv \dot{q} \cdot p - L = (\dot{q} \cdot \partial/\partial\dot{q} - 1)L \qquad (1.18)$$

is defined, a linear Legendre transform[3] of the Lagrangian expressed in terms of q and p. The expression for the total differential

$$dH = \dot{q} \cdot dp - \partial L/\partial q \cdot dq \qquad (1.19)$$

follows by evoking the definition (1.17). In combination with the Euler–Lagrange equations (1.1), (1.19) implies that the *Hamilton canonical equations*

$$\dot{q} = \partial H/\partial p, \qquad \dot{p} = -\partial H/\partial q \qquad (1.20)$$

are satisfied by $q = q(t)$ and $p = p(t)$. Thus, for example, the Lagrangian (1.2) produces the Hamiltonian

$$H = \tfrac{1}{2}p \cdot p + V \qquad (1.21)$$

and associated canonical equations

$$\dot{q} = p, \qquad \dot{p} = -\partial V/\partial q.$$

From the point of view of Hamilton with his $2n$-first-order dynamical equations (1.20), a state of the system is realized at any instant of time by fixing the *canonical variables* q and p, the generalized coordinate n-tuple and the generalized momentum n-tuple.

We have seen that for the exceptional case $n = 1$ the Euler–Lagrange equation (1.1) does not require an essentially unique Lagrangian nor a specific equivalence class of Lagrangians related by transformations (1.4). It follows that the Hamilton formulation is not indicated uniquely for a dynamical system with $n = 1$. In fact, the Hamiltonian can be taken to be an arbitrary function of the quantity $(\tfrac{1}{2}\dot{q}^2 + V)$ for an $n = 1$

---

[3] For a discussion of the geometrical and analytical significance of Legendre transformations, see R. Courant and D. Hilbert, "Methods of Mathematical Physics II," p. 32. Wiley (Interscience), New York, 1962.

dynamical system described by the Newtonian Euler–Lagrange equation (1.3).[4] On the other hand, if n is greater than 1, all known Hamiltonians equal the physical energy of the dynamical system. Moreover, with passage to the quantum theory, it is essential to identify the Hamiltonian with the physical energy, the latter being a strictly additive (as well as conserved) quantity for the composite of two noninteracting dynamical systems, like the frequency ω = (physical energy)/$\hbar$ associated with the composite quantum stationary state for two noninteracting systems. Henceforth, we assume that the Hamiltonian equals the physical energy for dynamical systems with $n = 1$, as it does generally for dynamical systems with $n > 1$. Because of Eqs. (1.5) and (1.18), this assumption is equivalent to identifying the canonical energy (1.5) with the observed physical energy, the latter quantity being given by $(E'_{phy} + E''_{phy})$ for a composite dynamical system composed of two noninteracting dynamical systems with physical energies E'_{phy} and E''_{phy}, respectively.

The Lagrange formulation is not completely equivalent to the Hamilton formulation, and the Hamilton formulation provides a dynamical description for certain special systems precluded by the former. To see this, note that as a consequence of (1.18) and the first canonical equation (1.20), we have the Lagrangian given as a linear Legendre transform of the Hamiltonian,

$$L = p \cdot \dot{q} - H = (p \cdot \partial/\partial p - 1)H. \tag{1.22}$$

---

[4] The first integral of the Newtonian form of the Hamilton canonical equations

$$\ddot{q} + \frac{dV}{dq} = \frac{\partial H}{\partial p}\frac{\partial^2 H}{\partial q\,\partial p} - \frac{\partial H}{\partial q}\frac{\partial^2 H}{\partial p^2} + \frac{dV}{dq} = 0$$

is

$$H = \text{arbitrary } C^1 \text{ function of } \left(\frac{1}{2}\left(\frac{\partial H}{\partial p}\right)^2 + V\right),$$

as noted by P. Havas, *Nuovo Cimento Suppl.* **5**, 363, 1957; F. J. Kennedy, Jr. and E. H. Kerner, *Amer. J. Phys.* **33**, 463, 1965. A linear first-order partial differential equation equivalent to the nonlinear first integral is obtained by evoking the Legendre transformation (1.18) with $L = \hat{L} \equiv \hat{L}(V, \dot{q})$, the quantity $\dot{q}(\partial \hat{L}/\partial \dot{q}) - \hat{L}$ being equal to an arbitrary C^1 function of $(\frac{1}{2}\dot{q}^2 + V)$, from which one obtains the complete integral (1.9) by using linear methods.

It is evident that the "inverse" Hamiltonian-to-Lagrangian mapping (1.22) relinquishes dynamical effects associated with a term of first-order homogeneity in p in the physical energy. Thus, for example, there is *no* Lagrange formulation for a dynamical system described by the Hamiltonian $H = (p \cdot p)^{1/2}$ (appropriate in the $n = 3$ case for a free relativistic spinless particle of mass zero) because the Lagrangian (1.22) vanishes identically for $H = (p \cdot p)^{1/2}$, and the associated Euler–Lagrange equations (1.1) are devoid of content. On the other hand, the Legendre transform (1.18) maps dynamically equivalent Lagrangians (related by a transformation (1.4)) into the same Hamiltonian. If the Hamiltonian (1.18) vanishes identically for a certain Lagrangian (for example, those of the form $L = A(q)(\dot{q} \cdot \dot{q})^{1/2}$), we have

$$dH/dt = \dot{q} \cdot (-\partial L/\partial q + (d/dt)\, \partial L/\partial \dot{q}) \equiv 0;$$

hence, the associated Euler–Lagrange equations (1.1) are linearly dependent and not *deterministic*, in the sense of being sufficient to yield a unique coordinate n-tuple $q(t)$ for $t > 0$ subject to prescribed initial data $(q(0), \dot{q}(0))$. A Lagrangian for which the associated Hamiltonian vanishes identically is of no physical (and dubious mathematical) interest, whereas there are physically significant Hamiltonians for which the associated Lagrangians vanish identically.

Let us consider the Hamilton formulation correspondent of the action principle (1.16). In place of the action functional (1.12) associated with the Lagrange formulation, we must employ the *canonical variable action functional*

$$\bar{S} = \bar{S}[q(t),\, p(t)] \equiv \int_{t'}^{t''} (\dot{q} \cdot p - H(q, p))\, dt, \tag{1.23}$$

obtained by using (1.18) to eliminate the Lagrangian from (1.12) and defined on the domain of all real n-tuple functions $q(t)$ of continuity class C^1 that reduce to the prescribed boundary values $q(t') = q'$ and $q(t'') = q''$, fixed n-tuples, and $p(t)$ of continuity class C^0 (unconstrained at $t = t'$ and $t = t''$). The $2n$ first-order Hamilton canonical equations (1.20) are equivalent to the *canonical variable action principle*

$$\delta \bar{S}/\delta q(t) = 0, \qquad \delta \bar{S}/\delta p(t) = 0, \tag{1.24}$$

where the functional derivatives are taken according to the general definition in Appendix A with

$$\mathscr{F} = \{\sigma(t)\colon \sigma(t) \in C^1 \quad \text{for} \quad t' \leqslant t \leqslant t'', \sigma(t') = \sigma(t'') = 0\}$$

in the case of the $q(t)$ differentiation and with

$$\mathscr{F} = \{\sigma(t) \colon \sigma(t) \in C^0 \quad \text{for} \quad t' \leqslant t \leqslant t''\}$$

in the case of the $p(t)$ differentiation. With these definitions of functional differentiation, the canonical variable action principle (1.24) states that $\bar{S}$ has an extremum for a physical pair of n-tuple functions $q(t)$, $p(t)$. Because $q(t)$ and $p(t)$ play asymmetric roles in (1.23) and (1.24), the canonical variable action principle does not provide a logical starting point for a discussion of canonical transformations.

3. Poisson Formulation

A useful feature of the Hamilton canonical equations (1.20) is that they allow the time rate of change of any *observable* $f = f(q, p)$, defined here as an infinitely differentiable real 1-tuple function in the canonical variables, to be expressed directly in terms of q and p as

$$df/dt \equiv \dot{f} = [f, H] \tag{1.25}$$

with the *Poisson bracket* of two observables introduced as

$$[f, g] \equiv \partial f/\partial q \cdot \partial g/\partial p - \partial g/\partial q \cdot \partial f/\partial p. \tag{1.26}$$

For example, consider the observable

$$f = \tfrac{1}{2} q \cdot \alpha \cdot q + q \cdot \beta \cdot p + \tfrac{1}{2} p \cdot \gamma \cdot p + \xi \cdot q + \eta \cdot p + \omega, \tag{1.27}$$

where α, β, and γ denote constant real n^2-dyads with α and γ symmetrical, ξ and η denote constant real n-tuples, and ω denotes a constant real 1-tuple; then the Poisson bracket of the observable (1.27) and the Hamiltonian observable[5] (1.21) is the observable

$$[f, H] = q \cdot \alpha \cdot p + p \cdot \beta \cdot p + \xi \cdot p$$
$$- q \cdot \beta \cdot \partial V/\partial q - p \cdot \gamma \cdot \partial V/\partial q - \eta \cdot \partial V/\partial q.$$

---

[5] If not originally so, a Hamiltonian can always be modified to be infinitely differentiable without altering the observable dynamics of the system. For example, if the potential energy in (1.21) were originally prescribed as the Coulombic form

$$V = (\text{const})(q \cdot q)^{-1/2},$$

the modified potential energy

$$V = (\text{const})(q \cdot q)^{-1/2} (\exp -\varepsilon(q \cdot q)^{-1/2})$$

would provide an infinitely differentiable Hamiltonian with physically indistinguishable dynamics for sufficiently small positive constant ε.

In particular, the observable constants of the motion are quantities which have a zero Poisson bracket with the Hamiltonian.

The set of all observables is closed under ordinary addition and Poisson bracket combination, in the sense that both the sum and the Poisson bracket of any two observables are observables. Furthermore, it is easy to verify[6] that the Poisson bracket combination law (1.26), associating a new observable function of q and p with an ordered pair of observable functions of q and p, has all the properties required of a *Lie product*[7] binary operation, namely:

$$[f + g, h] = [f, h] + [g, h] \quad \text{(linearity)} \quad (1.28)$$

$$[f, g] = -[g, f] \quad \text{(antisymmetry)} \quad (1.29)$$

$$[f, [g, h]] + [g, [h, f]] + [h, [f, g]] = 0 \quad \text{(integrability).} \quad (1.30)$$

Hence, the set of all observables subject to ordinary addition and Poisson bracket combination constitutes a *Lie ring* or *Lie algebra*.[8] The Poisson bracket (1.26) also has some special properties which are not generally featured by the Lie product in an abstract Lie algebra because two observables f and g can also be combined by ordinary multiplication to give the observable $fg = gf$, a quantity which is *not* generally found in a Lie algebra with elements $\{f, g, [f, g], \ldots\}$. Thus, for example, the Poisson bracket has the special property

$$[f, gh] = [f, g]h + g[f, h]. \quad (1.31)$$

The set of all observables subject to ordinary addition, ordinary multiplication, and Poisson bracket combination constitutes the *enveloping algebra* of the Lie algebra of observables.

Any subset of observables that is closed with respect to ordinary addition and Poisson bracket combination also constitutes a Lie algebra, more precisely, a Lie subalgebra of the Lie algebra of all observables.

---

[6] For proof of (1.30), see H. Goldstein, "Classical Mechanics," p. 256. Addison–Wesley, Cambridge, Massachusetts, 1953.

[7] For example, M. Hall, Jr., "Theory of Groups," p. 328. Macmillan, New York, 1959.

[8] See, for example, C. Chevalley, "Theory of Lie Groups," p. 103. Princeton Univ. Press, Princeton, New Jersey, 1946. The prototype for a Lie algebra is the set of all linear combinations of the generators G_i for a linear representation of a Lie group (see Appendix B) with the Lie product described as the commutator of two elements.

The subset of observables having the generic form (1.27), with elements in the subset labeled by fixing the $\frac{1}{2}n(n + 1)$ real parameters in α, the n^2 in β, the $\frac{1}{2}n(n + 1)$ in γ, the n in ξ, the n in η, and the 1 real parameter ω, constitutes a $(2n^2 + 3n + 1)$-dimensional Lie algebra, where "dimension" in this context is understood to mean the number of independent continuous real parameters required to label all elements in the Lie algebra.[9] Because this $(2n^2 + 3n + 1)$-dimensional Lie algebra contains those observables which are generally of prime physical significance (position, momentum, angular momentum, and so forth), we denote the subset of observables having the generic form (1.27) by a symbol, $\mathscr{A}_s$. The main structure of $\mathscr{A}_s$ is evident in the subset with $\xi = \eta = 0$, $\omega = 0$; this subset of observables, which we denote by $\mathscr{A}_s'$, is closed with respect to ordinary addition and Poisson bracket combination, and hence constitutes a $(2n^2 + n)$-dimensional Lie algebra. Finally, let us denote the subset of $\mathscr{A}_s$ with $\alpha = \beta = \gamma = 0$ by the symbol $\mathscr{A}_s''$; since this subset is also closed with respect to addition and Poisson bracket combination, it is a $(2n + 1)$-dimensional Lie algebra. Because the Poisson bracket of an observable in $\mathscr{A}_s'$ and an observable in $\mathscr{A}_s''$ is an observable in $\mathscr{A}_s''$, $\mathscr{A}_s$ is the *semidirect sum*[10] of $\mathscr{A}_s'$ and $\mathscr{A}_s''$,

$$\mathscr{A}_s = \mathscr{A}_s' \oplus \mathscr{A}_s''.$$

We note that $\mathscr{A}_s'$ is isomorphic to the well-known Lie algebra for the symplectic group in $2n$ dimensions, $Sp(2n)$.[11]

To the Lie algebra $\mathscr{A}$ of a subset of observables that is closed with respect to ordinary addition and Poisson bracket combination there is associated a conjugate Lie algebra $\mathscr{A}^*$ consisting of all observables which have a zero Poisson bracket with all elements of $\mathscr{A}$. We verify

---

[9] By definition, the Lie algebra of *all* observables is infinite-dimensional.

[10] In the present context, the term *semidirect sum* means that the Lie group associated with $\mathscr{A}$ is the semidirect product of the Lie groups associated with $\mathscr{A}'$ and $\mathscr{A}''$; for the classical definition of the semidirect product of groups, see M. Hall, Jr., "Theory of Groups," p. 88. Macmillan, New York, 1959.

[11] For a discussion of the symplectic groups, see M. Hamermesh, "Group Theory and Its Application to Physical Problems," pp. 402–412. Addison–Wesley, Reading, Massachusetts, 1962. We encounter a $2n$-dimensional matrix representation of $Sp(2n)$ if *linear* canonical transformations are considered. Using the notation of Eq. (1.48) and putting $Y_a = T_{ab} X_b$ with (T_{ab}) a constant $2n \times 2n$ array requires (T_{ab}) to satisfy the matrix equation $T_{ac} T_{bd} \Omega_{cd} = \Omega_{ab}$, and so the (T_{ab}) admissible for linear canonical transformations constitute a $2n$-dimensional representation of the $(2n^2 + n)$-parameter Lie group $Sp(2n)$.

that the subset $\mathscr{A}^*$ is, indeed, a Lie algebra by observing that if f and g are in $\mathscr{A}^*$,

$$[f, h] = [g, h] = 0 \qquad \text{for all} \quad h \in \mathscr{A}, \tag{1.32}$$

then

$$[f + g, h] = 0 \qquad \text{for all} \quad h \in \mathscr{A}, \tag{1.33}$$

because of (1.28) and (1.32), and

$$[[f, g], h] = [g, [h, f]] + [f, [g, h]]$$
$$= 0 \qquad \text{for all} \quad h \in \mathscr{A}, \tag{1.34}$$

because of (1.29), (1.30), and (1.32); therefore, the subset $\mathscr{A}^*$, closed with respect to ordinary addition and Poisson bracket combination, constitutes a Lie algebra. Such a Lie algebra $\mathscr{A}^*$ is infinite-dimensional because arbitrary (infinitely differentiable) functions of observables in $\mathscr{A}^*$ are also contained in $\mathscr{A}^*$. The observable constants of the motion constitute the Lie algebra

$$\mathscr{A}_H{}^* \equiv \{\phi(f_1, \ldots, f_{2n-1}) : \phi \in C^\infty, [f_i, H] = 0\}$$

conjugate to the 1-dimensional Lie algebra consisting of elements proportional to the Hamiltonian, $\mathscr{A}_H \equiv \{(\text{real constant parameter}) \times H\}$ because the general theory of linear first-order partial differential equations guarantees $(2n - 1)$ *functionally* independent solutions to $[f_i, H] = 0$. Associated with the infinite-dimensional Lie algebra $\mathscr{A}_H{}^*$, we have the finite-dimensional Lie algebra

$$\hat{\mathscr{A}}_H{}^* \equiv \left\{ \left(\sum_{i=1}^m \alpha_i w_i \right) : [w_i, H] = 0 \right\},$$

where the α's are real constants and the w's (assumed to be functionally independent of H) constitute a maximal finite set, closed under Poisson bracket combination, of linearly independent constants of the motion (exclusive of H). For example, the Hamiltonian

$$H = \tfrac{1}{2} p \cdot p = \tfrac{1}{2}(p_1{}^2 + p_2{}^2 + p_3{}^2)$$

with $n = 3$ yields

$$f_i = p_i, \qquad f_4 = q_2 p_3 - q_3 p_2, \qquad f_5 = q_3 p_1 - q_1 p_3$$

for the infinite-dimensional Lie algebra $\mathscr{A}_H{}^*$, and

$$w_i = p_i, \qquad w_4 = q_2 p_3 - q_3 p_2, \qquad w_5 = q_3 p_1 - q_1 p_3,$$
$$w_6 = q_1 p_2 - q_2 p_1$$

for the associated 6-dimensional $\mathscr{A}_H{}^*$; here, the constant of the motion

$$w_6 = -(w_1 w_4/w_3) - (w_2 w_5/w_3)$$

is *linearly* independent of the other w's but functionally dependent on them.

A transformation of all observables

$$f = f(q, p) \overset{w}{\to} f_\tau = f_\tau(q, p)$$

$$g = g(q, p) \overset{w}{\to} g_\tau = g_\tau(q, p) \tag{1.35}$$

$$\cdots$$

such that

$$f_0 = f, \qquad g_0 = g, \qquad \cdots \tag{1.36}$$

is said to be a *contact* or *canonical transformation* generated by the observable $w = w(q, p)$ if the transformed observables satisfy the first-order total differential equations

$$df_\tau/d\tau = [f_\tau, w]$$

$$dg_\tau/d\tau = [g_\tau, w] \tag{1.37}$$

$$\cdots\cdots$$

A transformed observable is given explicitly in terms of the original observable by the Maclaurin series derived from (1.36) and (1.37),

$$f_\tau = f + \tau[f, w] + \frac{\tau^2}{2!}[[f, w], w] + \cdots$$

$$\equiv \left(\exp \tau\left(\frac{\partial w}{\partial p} \cdot \frac{\partial}{\partial q} - \frac{\partial w}{\partial q} \cdot \frac{\partial}{\partial p}\right)\right)f. \tag{1.38}$$

Also note that we have

$$h = [f, g] \Rightarrow h_\tau = [f_\tau, g_\tau] \tag{1.39}$$

for $\tau \geqslant 0$, as readily verified by direct computation:

$$\frac{d}{d\tau}(h_\tau - [f_\tau, g_\tau]) = \frac{dh_\tau}{d\tau} - \left[\frac{df_\tau}{d\tau}, g_\tau\right]$$

$$- \left[f_\tau, \frac{dg_\tau}{d\tau}\right] = [h_\tau, w] - [[f_\tau, w], g_\tau] \tag{1.40}$$

$$- [f_\tau, [g_\tau, w]] \equiv [(h_\tau - [f_\tau, g_\tau]), w].$$

Because of property (1.39), canonical transformations preserve the Lie algebraic structure for a subset of observables that is closed with respect to addition and Poisson bracket combination: Lie algebras are mapped into isomorphic Lie algebras by canonical transformations. Moreover, the time rate of change of any canonically transformed observable is prescribed by a dynamical equation of the form (1.25),

$$df_\tau/dt \equiv \dot{f}_\tau = [f_\tau, H_\tau], \qquad (1.41)$$

as seen by putting $h = f$ and $g = H$ in (1.39). Next, by specializing (1.39) to the cases for which $h = c$, a numerical constant independent of q and p, we note that (1.38) produces $h_\tau = c$ for $\tau \geqslant 0$ because $[c, w] = 0$ for all w, and hence

$$[f, g] = c \Rightarrow [f_\tau, g_\tau] = c. \qquad (1.42)$$

In particular, for the components of the generalized coordinate n-tuple and generalized momentum n-tuple, we have the basic Poisson bracket relations

$$[q_i, q_j] = [p_i, p_j] = 0, \qquad [q_i, p_j] = \delta_{ij}, \qquad (1.43)$$

so that (1.42) yields

$$[q_{\tau i}, q_{\tau j}] = [p_{\tau i}, p_{\tau j}] = 0, \qquad [q_{\tau i}, p_{\tau j}] = \delta_{ij}. \qquad (1.44)$$

It is an important consequence of (1.44) that if two observables f and g are regarded as functions of the canonically transformed components of the generalized coordinate and momentum n-tuples, their Poisson bracket computed with respect to q_τ and p_τ equals their Poisson bracket computed with respect to q and p,

$$\frac{\partial f}{\partial q_\tau} \cdot \frac{\partial g}{\partial p_\tau} - \frac{\partial g}{\partial q_\tau} \cdot \frac{\partial f}{\partial p_\tau} = \frac{\partial f}{\partial q} \cdot \frac{\partial g}{\partial p} - \frac{\partial g}{\partial q} \cdot \frac{\partial f}{\partial p}. \qquad (1.45)$$

To prove (1.45) in a lucid manner, it is expedient to introduce some extra notation. Let $X_i = q_i$ and $X_{i+n} = p_i$ for $i = 1$ to n and define the $2n \times 2n$ array of constants

$$\Omega_{ab} \equiv \begin{array}{ll} 1 & \text{for} \quad a = b - n \\ -1 & \text{for} \quad a = b + n \\ 0 & \text{otherwise.} \end{array} \qquad (1.46)$$

Then the Poisson bracket of two observables can be expressed as

$$[f, g] = \frac{\partial f}{\partial X_a} \Omega_{ab} \frac{\partial g}{\partial X_b}, \qquad (1.47)$$

where the repeated indices are understood to be summed from 1 to $2n$. Let $Y_i = q_{\tau i}$ and $Y_{i+n} = p_{\tau i}$ for $i = 1$ to n. Then the content of Eqs. (1.44) can be expressed as[12]

$$\frac{\partial Y_a}{\partial X_c} \Omega_{cd} \frac{\partial Y_b}{\partial X_d} = \Omega_{ab}, \tag{1.48}$$

from which it follows immediately that

$$\frac{\partial f}{\partial X_a} \Omega_{ab} \frac{\partial g}{\partial X_b} = \frac{\partial f}{\partial Y_c} \frac{\partial Y_c}{\partial X_a} \Omega_{ab} \frac{\partial Y_d}{\partial X_b} \frac{\partial g}{\partial Y_d} = \frac{\partial f}{\partial Y_c} \Omega_{cd} \frac{\partial g}{\partial Y_d}. \tag{1.49}$$

Passing back to canonical variable notation, one obtains (1.45) from (1.49).

Because of Eq. (1.45), the Poisson bracket in an equation can be computed with respect to q_τ, p_τ for any value of $\tau \geqslant 0$. By doing this in (1.41) and specializing to the cases for which f_τ equals each of the $2n$ components of q_τ and p_τ, we obtain Hamilton canonical equations for the transformed quantities,

$$\dot{q}_\tau = \partial H_\tau / \partial p_\tau, \qquad \dot{p}_\tau = -\partial H_\tau / \partial q_\tau. \tag{1.50}$$

Hence, canonical transformations preserve the form of the Hamilton equations, as well as the form of the Poisson brackets.

The notion of canonical transformations generated by an observable w can be generalized to the notion of canonical transformations generated by a finite-dimensional Lie algebra. Consider an m-dimensional Lie algebra $\mathscr{A}$ composed of all real linear combinations of m linearly independent observables $w_1, \ldots, w_m$. Since $\mathscr{A}$ is closed with respect to Poisson bracket combination, we must have

$$[w_i, w_j] = -\sum_{k=1}^{m} c_{ijk} w_k, \tag{1.51}$$

---

[12] In language familiar to students of general relativity, Eq. (1.48) states that the skew-symmetric tensor Ω is invariant under canonical point transformations $X \to Y = Y(X; \tau)$ in the $2n$-dimensional *phase space*. Having established the existence of the nonsingular invariant tensor Ω, it is easy to derive the canonical transformation invariants of Poincaré. For instance, we find

$$\prod_{a=1}^{2n} dX_a = \prod_{a=1}^{2n} dY_a,$$

that is,

$$\prod_{i=1}^{n} dq_i \, dp_i = \prod_{i=1}^{n} dq_{\tau i} \, dp_{\tau i},$$

by taking the determinant of (1.48) and evoking connectivity to the identity transformation to get $\det(\partial Y_a / \partial X_b) = +1$.

where the $c_{ijk} = -c_{jik}$ are certain real constants, the *structure constants* of the Lie algebra. Note that it follows from (1.30) and the linear independence of the w's that the structure constants in (1.51) satisfy the quadratic *Lie identities*,

$$\sum_{h=1}^{m} (c_{ijh} c_{hkl} + c_{jkh} c_{hil} + c_{kih} c_{hjl}) \equiv 0. \tag{1.52}$$

We introduce the *generators* for canonical transformations as

$$G_i \equiv \frac{\partial w_i}{\partial p} \cdot \frac{\partial}{\partial q} - \frac{\partial w_i}{\partial q} \cdot \frac{\partial}{\partial p}, \tag{1.53}$$

the m first-order partial differential operators that satisfy the commutation relations

$$G_i G_j - G_j G_i = \sum_{k=1}^{m} c_{ijk} G_k \tag{1.54}$$

as a consequence of the Poisson bracket closure relations (1.51). A linear representation of the m-parameter Lie group (see Appendix B) associated with the Lie algebra $\mathscr{A}$ is provided by linear differential operators of the form $(\exp \alpha \cdot G) \equiv \sum_{N=0}^{\infty} (N!)^{-1} (\alpha \cdot G)^N$, parametrized by a real m-tuple $\alpha = (\alpha_1, \ldots, \alpha_m)$. Under canonical transformations generated by $\mathscr{A}$, a generic observable $f = f(q, p)$ is transformed to

$$f_\alpha = (\exp \alpha \cdot G)f, \tag{1.55}$$

as seen by putting $\tau = 1$ and $w = \sum_{i=1}^{m} \alpha_i w_i$ in (1.38). Conversely, by setting $\alpha = \tau \beta$ and $\sum_{i=1}^{m} \beta_i w_i = w$ with β a fixed unit m-tuple, $\beta \cdot \beta \equiv 1$, we obtain (1.38) from (1.55). It follows that the m-parameter canonical transformations generated by $\mathscr{A}$ produce transformed observables (1.55) with all the properties derived for the transformed observables (1.38). For example, the m-parameter canonical transformation generalization of (1.39) is

$$h = [f, g] \Rightarrow h_\alpha = [f_\alpha, g_\alpha], \tag{1.56}$$

and the implied generalization of (1.41) is

$$df_\alpha / dt \equiv \dot{f}_\alpha = [f_\alpha, H_\alpha]. \tag{1.57}$$

As an illustration of canonical transformations generated by a finite-dimensional Lie algebra, consider the three-dimensional subalgebra of $\mathscr{A}_s$,

$$\mathscr{A} = \{\alpha_1 w_1 + \alpha_2 w_2 + \alpha_3 w_3 : w_1 \equiv \tfrac{1}{4}(p \cdot p - q \cdot q),$$
$$w_2 \equiv -\tfrac{1}{4}(q \cdot q + p \cdot p), \quad w_3 \equiv -\tfrac{1}{2}q \cdot p\}. \tag{1.58}$$

For this $\mathscr{A}$ the Poisson bracket closure relations (1.51) are

$$[w_1, w_2] = -w_3, \qquad [w_2, w_3] = -w_1, \qquad [w_3, w_1] = w_2. \quad (1.59)$$

Hence, the independent nonzero structure constants are

$$c_{123} = c_{231} = -c_{312} = 1.$$

The Lie algebra (1.58) is isomorphic to the Lie algebra for $SL(2, R)$ (see Appendix B). The associated generators (1.53),

$$G_1 = \frac{1}{2}\left(q \cdot \frac{\partial}{\partial p} + p \cdot \frac{\partial}{\partial q}\right), \qquad G_2 = \frac{1}{2}\left(q \cdot \frac{\partial}{\partial p} - p \cdot \frac{\partial}{\partial q}\right)$$

$$G_3 = \frac{1}{2}\left(p \cdot \frac{\partial}{\partial p} - q \cdot \frac{\partial}{\partial q}\right) \qquad (1.60)$$

give a representation of $SL(2, R)$ canonical transformations on the space of observables $f = f(q, p)$.

Special importance is attached to canonical transformations generated by the Hamiltonian and by the simpler Lie algebras associated with the Hamiltonian. First, by putting $\tau = t$, $f_t = f_t(q(0), p(0)) \equiv f(q(t), p(t))$, and $w = H(q(0), p(0))$ in (1.37), it is evident that the dynamical evolution of all observables, as prescribed by (1.25), can be viewed as a canonical transformation parametrized by time and generated by the Hamiltonian. This special canonical transformation takes $q(0) \overset{H}{\to} q_t(0) \equiv q(t)$ and $p(0) \overset{H}{\to} p_t(0) \equiv p(t)$, with the Poisson bracket in (1.37) computed most conveniently with respect to the "original" canonical variables $(q(0), p(0))$. The dynamical evolution of all observables is expressed in the canonical transformation form (1.38) as

$$f = f(q(t), p(t)) = \left(\exp t\left(\frac{\partial H}{\partial p(0)} \cdot \frac{\partial}{\partial q(0)} - \frac{\partial H}{\partial q(0)} \cdot \frac{\partial}{\partial p(0)}\right)\right)f(q(0), p(0)). \quad (1.61)$$

Canonical transformations generated by the Lie algebra $\mathscr{A}_H{}^*$ are especially noteworthy since (1.55) shows that the Hamiltonian is invariant for them,

$$H_\alpha \equiv H \qquad \text{for} \quad \hat{\mathscr{A}}_H{}^* \equiv \{\alpha \cdot w : [H, w_i] = 0\}, \qquad (1.62)$$

and (1.57) becomes

$$\dot{f}_\alpha = [f_\alpha, H]. \qquad (1.63)$$

The latter is a generic dynamical equation identical to (1.25), and thus the dynamics of observables is unchanged. For this reason, the Lie algebra $\hat{\mathscr{A}}_H{}^*$ is referred to as the *symmetry algebra* for the dynamical system, and the associated Lie group $\hat{\mathscr{G}}_H{}^*$, with its linear representation on the space of observables $\hat{\mathscr{G}}_H{}^* = \{(\exp \alpha \cdot G)\}$ provided by (1.53), is called the *symmetry group*. Finally, we note the especially simple kind of dynamics that arises if H is an element of the Lie algebra $\mathscr{A}_s$ discussed above, that is, if H has the form of the right-hand side of (1.27). Because the components of q and p are elements of $\mathscr{A}_s{}''$ and their Poisson brackets with elements of $\mathscr{A}_s$ are elements of $\mathscr{A}_s{}''$, it follows from (1.55) that $\mathscr{A}_s$ generates canonical transformations for which the components of q_α are inhomogeneous linear expressions in the components of q and p, and likewise for the components of p_α. In particular, the dynamical evolution generated by an H in $\mathscr{A}_s$ is characterized by the components of $q(t)$ being inhomogeneous linear expressions in the components of $q(0)$ and $p(0)$, and likewise for the components of $p(t)$. Moreover, if H is an element of $\mathscr{A}_s$, all observables in $\mathscr{A}_s$ at $t = 0$ remain in $\mathscr{A}_s$ for $t > 0$, and any observable that is not initially in $\mathscr{A}_s$ does enter the algebra, the constant of the motion H being the "fixed point" in $\mathscr{A}_s$ required by topological considerations. Dynamical containment in $\mathscr{A}_s$ is featured by all elements in the Lie algebra if H is in $\mathscr{A}_s$,

$$f(q(t), p(t)) \equiv f_t(q(0), p(0)) = \tfrac{1}{2}q(0) \cdot \alpha(t) \cdot q(0)$$
$$+ q(0) \cdot \beta(t) \cdot p(0) + \tfrac{1}{2}p(0) \cdot \gamma(t) \cdot p(0)$$
$$+ \xi(t) \cdot q(0) + \eta(t) \cdot p(0) + \omega(t), \tag{1.64}$$

because of the Lie algebra addition and Poisson bracket closure properties.

Chapter 2 *Quantum Mechanics*

1. Feynman Formulation

Again taken as a primitive physical notion, the *state* of a dynamical system with a finite number of degrees of freedom is realized at any instant of time by prescribing the complex *wave function* $\psi = \psi(q)$ (function class C^1, piecewise C^2, with respect to $q = (q_1, \ldots, q_n)$ and absolute-square-integrable over all q). The associated positive-definite quantity $|\psi(q)|^2$ is interpreted as the relative *probability density* for a physical measurement locating the state at the generalized coordinate n-tuple q. Contrasting sharply with the mathematical realization of the state in classical theory, this role assigned to a complex wave function is the basic postulate in quantum mechanics.

The system *dynamics* is described by assigning the wave function a time-dependent form $\psi = \psi(q; t)$ (C^1 with respect to t). A physically admissible evolution in time takes the specific *linear* form of a *dynamical principle of superposition*

$$\psi(q''; t'') = \int K(q'', q'; t'' - t')\psi(q'; t')\, dq' \tag{2.1}$$

for all $t'' \geqslant t'$ and initial wave functions $\psi = \psi(q'; t')$, where the integration is over all q'; the q'-space infinitesimal volume element, abbreviated as $dq' \equiv$ (function of q') $\prod_{i=1}^{n} dq_i'$, reduces to $dq' = \prod_{i=1}^{n} dq_i'$ for the

22

special case of linear coordinates. Note that the *propagation kernel* $K(q'', q'; t'' - t')$ in (2.1) must satisfy a *semigroup composition law*[1]

$$K(q'', q'; t'' - t') = \int K(q'', q; t'' - t)K(q, q'; t - t') \, dq \qquad (2.2)$$

for $t' \leqslant t \leqslant t''$, as well as the initial value condition

$$\lim_{t \to 0} K(q'', q'; t) = \delta(q'' - q'), \qquad (2.3)$$

in order to guarantee the general consistency of (2.1) for all $\psi(q'; t')$. Repeated application of (2.2) shows that the propagation kernel for finite values of $(t'' - t')$ can be expressed as an iteration of the propagation kernel for infinitesimal values of $(t'' - t')$, by letting N increase without bound in the equation

$$K(q'', q'; t'' - t') = \int \left(\prod_{M=1}^{N} K(q^{(M)}, q^{(M-1)}; \Delta t) \right) \prod_{M=1}^{N-1} dq^{(M)}$$

$$\Delta t \equiv (t'' - t')/N, \qquad q^{(0)} \equiv q', \qquad q^{(N)} \equiv q''. \quad (2.4)$$

Before letting N increase without bound in (2.4), an obvious notational simplification is to introduce an *n*-tuple function $q(t)$ that equals $q^{(M)}$ for each integer value of $(t - t')/\Delta t \equiv M$ from 0 to N,

$$q(t' + M \, \Delta t) \equiv q^{(M)}, \qquad (2.5)$$

[1] For the abstract definition of a semigroup, see M. Hall, Jr., "Theory of Groups," p. 7. Macmillan, New York, 1959. In the present context, the one-parameter set of propagation kernels $K(q'', q'; t)$, with elements in the set labeled by all real $t \geqslant 0$, constitutes a Lie semigroup, the associative binary product operation being given by (2.2) and the identity element being prescribed consistently by (2.3). In order to conserve probability, it is necessary and sufficient to have

$$\int K(q, q''; t)^* K(q, q'; t) \, dq = \delta(q'' - q')$$

for $t > 0$. If the propagation kernel manifests the latter property, it is possible to extend the semigroup and turn it into a one-parameter Lie group by defining the propagation kernel for negative values of t by complex-conjugation, $K(q'', q'; -t) \equiv K(q', q''; t)^*$, and requiring (2.2) to hold for all real t', t'', and t. This definition of the propagation kernel for negative values of t is quite natural since it is the analytic continuation of the Feynman functional integral representation (2.8) if the normalization constant in (2.7) is such that $A(\Delta t)^* = A(-\Delta t)$, as in the case of a Lagrangian of the form (1.2). Notwithstanding the natural quality of the definition $K(q'', q'; -t) \equiv K(q', q''; t)^*$, it has become customary to define $K(q'', q'; t) \equiv 0$ for $t < 0$ to simplify the formal appearance of perturbation theory (for example, see R. P. Feynman, "Quantum Electrodynamics," p. 72. Benjamin, New York, 1961).

with $q^{(0)} = q' = q(t')$ and $q^{(N)} = q'' = q(t'')$, the fixed n-tuple arguments in $K(q'', q'; t'' - t')$. By introducing this notation into (2.4), we have

$$K(q'', q'; t'' - t')$$

$$= \int \left(\prod_{M=1}^{N} K(q(t' + M \, \Delta t), q(t' + (M - 1) \, \Delta t); \Delta t) \right) \prod_{M=1}^{N-1} dq(t' + M \, \Delta t).$$

$$(2.6)$$

The advantage of the parametrization in (2.6) is that in the limit $N \to \infty$, $\Delta t \to 0$, the time variable in $q(t)$ serves as a virtually continuous parameter for the intermediate integrations between t' and t''.

Equation (2.6) holds identically for any propagation kernel satisfying (2.2), and in order to make the dynamical principle of superposition (2.1) a definite physical law, we must prescribe the infinitesimal propagation kernel $K(q(t + \Delta t), q(t); \Delta t)$ in Eq. (2.6). On the basis of considerations by Dirac,[2] Feynman[3] postulated the infinitesimal propagation kernel to be given by

$$K(q(t + \Delta t), q(t); \Delta t) = A^{-1} \exp(iL([q(t + \Delta t) + q(t)]/2,$$

$$[q(t + \Delta t) - q(t)]/\Delta t) \, \Delta t/\hbar), \qquad (2.7)$$

where the normalization constant $A = A(\Delta t)$ depends on the form of the Lagrangian $L(q, \dot{q})$ for the dynamical system in such a way that

$$\lim_{\Delta t \to 0} K(q'', q'; \Delta t) = \delta(q'' - q')$$

(for example, $A = (2\pi i \hbar \, \Delta t)^{n/2}$ in the case of a Lagrangian of the form (1.2)), $\hbar$ is Planck's constant divided by 2π, and terms of the order $(\Delta t)^2$ are neglected. By putting (2.7) into (2.6) and taking the limit $N \to \infty$, $\Delta t \to 0$, the Feynman *functional integral representation* for the propagation kernel is obtained,

$$K(q'', q'; t'' - t') = \int_{\mathscr{C}} (\exp iS/\hbar) D(q), \qquad (2.8)$$

[2] P. A. M. Dirac, "The Principles of Quantum Mechanics," pp. 125–130. Oxford Univ. Press, New York and London, 1947.

[3] R. P. Feynman, Princeton Ph.D. Dissertation, 1942; *Rev. Modern Phys.* **20**, 367, 1948; also see R. P. Feynman and A. R. Hibbs, "Quantum Mechanics and Path Integrals." McGraw-Hill, New York, 1965.

where the action functional $S = S[q(t)]$ defined by (1.3) appears in the form of a Riemann sum,

$$S = \lim_{\Delta t \to 0} \sum_{M=1}^{N=(t''-t')/\Delta t} L([q(t' + M\,\Delta t) + q(t' + (M-1)\,\Delta t)]/2,$$

$$[q(t' + M\,\Delta t) - q(t' + (M-1)\,\Delta t)]/\Delta t)\,\Delta t, \quad (2.9)$$

and $D(q)$ in (2.8) denotes an infinitesimal volume element for the integration over the class of real n-tuple functions $\mathscr{C} = \{q = q(t)$ for $t' \leqslant t \leqslant t'' : q(t') = q', q(t'') = q''\}$. In symbolic notation, we have

$$D(q) = \mathcal{N} \prod_{t' < t < t''} dq(t) \qquad (2.10)$$

with t taking on all real values between t' and t'' in the infinite product of q-space infinitesimal volume elements and the generic normalization factor

$$\mathcal{N} = \mathcal{N}(t'' - t') \equiv \lim_{\Delta t \to 0} [A(\Delta t)]^{-(t''-t')/\Delta t} \qquad (2.11)$$

independent of $q(t)$.

The Feynman functional integral (2.8) bears a formal relationship to the Wiener functional integral (see Appendix D). This relationship has been exploited to put the existence of (2.8) on a sound mathematical basis.[4] There is, however, a more direct way to establish a meaning for the Feynman functional integral (2.8), without recourse to the Wiener functional integral. In the following we interpret the infinitesimal volume element (2.10) to be a Haar measure,[5] like the Haar measures for linear representations of Lie groups discussed in Appendix C.

In order to arrive at a meaning for the Feynman functional integral (2.8), it is expedient to work with a linear coordinate n tuple, so that $dq(t) = \prod_{i=1}^{n} dq_i(t)$. Then by introducing the real n-tuple function

$$\hat{q} = \hat{q}(t) \equiv q(t) - [((t - t')q'' + (t'' - t)q')/(t'' - t')], \qquad (2.12)$$

---

[4] R. H. Cameron, *J. Math. and Phys.* **39**, 126 (1960); J. A. Beekman, *J. Math. and Phys.* **46**, 253 (1967). For a recent review of Feynman–Wiener integration theory, see L. Streit, *Acta Phys. Austriaca (Suppl.)* **2**, 2 (1965).

[5] More precisely, $D(q)$ is the limit of a sequence of Haar measures; in calling $D(q)$ a Haar measure, we exercise the same physical license that is involved in calling $\delta(x)$ a δ-"function." Haar measure was evoked for the Feynman quantization of general relativity by C. W. Misner, *Rev. Modern Phys.* **29**, 497 (1957).

we have $dq(t) = d\hat{q}(t)$ for a fixed value of t since the square-bracketed term in (2.12) is constant for t fixed. Hence, (2.10) yields $D(q) = D(\hat{q})$ and (2.8) becomes

$$K(q'', q'; t'' - t') = \int_{\mathscr{G}} (\exp iS/\hbar) D(\hat{q}) \qquad (2.13)$$

with the integration over the space of n-tuple functions

$$\mathscr{G} = \{\hat{q} = \hat{q}(t) \quad \text{for} \quad t' \leqslant t \leqslant t'' : \hat{q}(t') = 0 = \hat{q}(t'')\}.$$

Observe that the infinitesimal volume element $D(\hat{q})$ is *displacement-invariant* for any fixed σ in $\mathscr{G}$,

$$D(\hat{q} + \sigma) = D(\hat{q}), \qquad (2.14)$$

because

$$D(\hat{q} + \sigma) = \mathscr{N} \prod_{t' < t < t''} d(\hat{q}(t) + \sigma(t)) = \mathscr{N} \prod_{t' < t < t''} d\hat{q}(t).$$

Moreover, the space of n-tuple functions $\mathscr{G}$ has the following properties:

1. It admits a topology with the *distance* between any pair of elements $u, v \in \mathscr{G}$ defined as

$$\|u - v\| \equiv \max_t \left[(u(t) - v(t)) \cdot (u(t) - v(t)) \right]^{1/2}.$$

2. It is an Abelian group under addition.

3. In the topology with the distance between elements u, v defined as $\|u - v\|$, finite-dimensional subgroups of $\mathscr{G}$ are locally compact, since every finite-dimensional normed linear space is locally compact; thus, a finite-dimensional subgroup of $\mathscr{G}$ is a continuous Abelian *topological group*.[6]

These properties suggest that the integral (2.13) be considered as the limit $m \to \infty$ of the integral of $(\exp iS/\hbar)$ over an m-dimensional Abelian subgroup of $\mathscr{G}$. Because the infinitesimal volume element is displacement-invariant, it is a left-invariant and right-invariant *Haar measure*[7] for the integration over an m-dimensional subgroup of $\mathscr{G}$.

---

[6] For example, J. L. Kelley, "General Topology," p. 105. Van Nostrand, Princeton, New Jersey, 1955.

[7] For example, P. R. Halmos, "Measure Theory," pp. 250–289. Van Nostrand, Princeton, New Jersey, 1950.

Displacement-invariance of the Haar measure defines this mth approximation to the propagation kernel uniquely to within a normalization factor independent of q'' and q' by virtue of the general Haar measure existence and uniqueness theorems.[7] The limiting form of the infinitesimal volume element as $m \to \infty$, denoted by $D(\hat{q})$ in (2.13), is not a Haar measure in the mathematical sense, since an infinite-dimensional normed linear space is not locally compact. However, for practical applications in physics, m need only be taken sufficiently large to yield requisite accuracy in the propagation kernel. Viewed physically in the approximation of a large but finite m, $D(\hat{q})$ is a Haar measure. The essential form of the normalization factor is fixed in practice by the physically required existence of the propagation kernel (2.13) for all $t'' \geqslant t'$, by the semigroup composition law (2.2), and by the initial-value condition (2.3). In general, an undetermined complex constant remains in $\mathcal{N}$, for (2.2) and (2.3) admit renormalization transformations of the form

$$\mathcal{N} \to [\exp z(t'' - t')]\mathcal{N}, \tag{2.15}$$

where z is an arbitrary complex constant; however, conservation of probability[1] requires the z in (2.15) to be pure imaginary, so that a renormalization transformation (2.15) has no observable physical consequences, corresponding to a translation of the energy reference scale [see, for example, Eq. (2.66)]. In the notation of Appendices A, B, and C, the Haar measure $D(\hat{q})$ is associated with the linear representation of $\mathcal{G}$

$$T(\hat{q}) \equiv \exp\left(\int_{t'}^{t''} \hat{q}(t) \cdot [\delta/\delta u(t)] \, dt\right)$$

on the space of infinitely differentiable *functionals* of the real n-tuple function $u = u(t) \in \mathcal{G}$. The Haar measure $D(\hat{q})$ [or, in the notation of Appendix C, $D(T(\hat{q}))$ with $m = \infty$] follows from this $T(\hat{q})$ just as the left-invariant and right-invariant Haar measure (C.16) follows to within normalization from the linear representation of the m-dimensional Abelian Lie group given by (B.12).

Equations (2.1) and (2.13) constitute the Feynman "sum-over-histories" formulation of the dynamical law for a system with a finite number of degrees of freedom. If we could evaluate the functional integral (2.13) for all physically important actions $S = S[q(t)]$, the theory would suffice in a very elegant way for the quantum mechanics of a closed nondissipative system with a classical Lagrangian, no Schrödinger or Dirac "abstract operator formalism" being necessary to describe the quantum mechanics of such a system. However, the functional integral representation of the propagation kernel (2.13) is not amenable to rigorous evaluation except for an action which is a *general quadratic* expression (that is, possibly including a linear functional term) in $q(t)$. If $S = S[q(t)]$ is a general quadratic, we have

$$S[q(t)] = S[q_c(t)] + S[\tilde{q}(t)] \tag{2.16}$$

where $\tilde{q}(t) \equiv q(t) - q_c(t)$ with $q_c(t)$ the classical "history" satisfying the action principle (1.16),

$$\delta S/\delta q(t)\Big|_{q=q_c} = 0, \tag{2.17}$$

subject to the boundary conditions $q_c(t') = q'$ and $q_c(t'') = q''$. With $q(t)$ an element of $\mathscr{C}$, it follows that the variable n-tuple function $\tilde{q}(t)$ is an element of $\mathscr{G}$, that is, $\tilde{q}(t') = \tilde{q}(t'') = 0$. Hence, by setting the fixed σ in (2.14) equal to

$$\sigma(t) = [((t - t')q'' + (t'' - t)q')/(t'' - t')] - q_c(t), \tag{2.18}$$

we find that the displacement-invariant measure $D(\hat{q}) = D(\tilde{q})$. Therefore, (2.13) produces the result

$$K(q'', q'; t'' - t') = \mathscr{N}(\exp iS[q_c(t)]/\hbar) \tag{2.19}$$

with the factor

$$\int_{\mathscr{G}} (\exp iS[\tilde{q}(t)]/\hbar) D(\tilde{q})$$

independent of q', q'' and absorbed into the normalization (2.11). As a prime example, the Lagrangian

$$L = \tfrac{1}{2}\left(\dot{q} \cdot \dot{q} - \sum_{i=1}^{n} \omega_i^2 q_i^2\right) \tag{2.20}$$

for a system of uncoupled simple harmonic oscillators leads to a propagation kernel of the form (2.19) with

$$\mathscr{N} = \prod_{i=1}^{n} \left(\frac{\omega_i}{2\pi i\hbar \sin \omega_i(t'' - t')}\right)^{1/2}$$

$$S[q_c(t)] = \sum_{i=1}^{n} \left[\frac{\omega_i(q_i''^2 + q_i'^2)}{2 \tan \omega_i(t'' - t')} - \frac{\omega_i q_i'' q_i'}{\sin \omega_i(t'' - t')}\right], \tag{2.21}$$

modulo a renormalization transformation (2.15). It is also noteworthy that the propagation kernel for the Lagrangian associated with free motion, $L = \frac{1}{2}\dot{q} \cdot \dot{q}$, is embraced by the limiting form of (2.21) with all

$$\omega_i \to 0: \quad \mathcal{N} = [2\pi i\hbar(t'' - t')]^{-n/2}$$

$$S[q_c(t)] = (q'' - q') \cdot (q'' - q')/2(t'' - t').$$

In spite of a very considerable research effort, the rigorous evaluation of (2.13) has not been carried through except for an action functional which is a general quadratic expression in $q(t)$.

There are several ways to proceed with the Feynman functional integral representation (2.13), explicit evaluation notwithstanding. First, we can extract useful dynamical information from (2.13) by asymptotic techniques and other suitable analytical procedures,[8] as is customary in the simpler, but analogous, case of ordinary integral representations for functions which cannot be evaluated explicitly. For example, if the action functional takes the form $S = S_0 + S_{int}$, where $S_0 = S_0[q(t)]$ is simple (that is, quadratic in $q(t)$) and $S_{int} = S_{int}[q(t)]$ is relatively small compared to S_0 for typical $q(t)$, then the propagation kernel can be evaluated approximately by working out the leading terms in the expansion of (2.13),

$$K(q'', q'; t'' - t') = \sum_{k=0}^{\infty} (k!)^{-1}(i/\hbar)^k \int_{\mathscr{G}} (S_{int})^k (\exp iS_0/\hbar)D(\hat{q}).$$

$$(2.22)$$

The latter perturbation expansion can be related directly to the conventional *iteration solution* for the wave function in the Schrödinger formulation (see Appendix H). Applications of (2.22) lead one to study functional integrals of the form

$$\int_{\mathscr{G}} A(\exp iS/\hbar)D(\hat{q}) \equiv (q''; t'' |A[q(t)]| q'; t')_s \qquad (2.23)$$

where $S = S[q(t)]$ and $A = A[q(t)]$ are generic real-valued functionals of $q(t)$; in the case of terms appearing in (2.22), we have $S = S_0$ and $A = (S_{int})^k$. Although all relevant and necessary dynamical information is in principle given by the functional integral for the propagation kernel (2.13), functional integrals of the form (2.23), sometimes called *Feynman*

---

[8] L. Streit, *Acta Phys. Austriaca (Suppl.)* **2**, 2 (1965); S. G. Brush, *Rev. Modern Phys.* **33**, 79 (1961); G. Rosen, *J. Math. Phys.* **4**, 1327 (1963); R. L. Zimmerman, *J. Math. Phys.* **6**, 1117 (1965); C. S. Lam, *Nuovo Cimento* **47**, 451 (1966).

operators, play the more prominent role in practical calculations. For all of the functional integrals in the theory with displacement-invariant Haar measure, there is a valuable "functional integration by parts" lemma, derived in Appendix D. This lemma allows one to express *exact* equations involving Feynman operators without having to do any explicit functional integration. For example, if we apply Eq. (D.1) to the propagation kernel itself by putting $F = (\exp iS/\hbar)$, we obtain

$$\int_{\mathcal{g}} [\delta S/\delta q(t)](\exp iS/\hbar)D(\hat{q}) = 0 \qquad \text{for} \quad t' < t < t'', \qquad (2.24)$$

which combined with the definition (2.23) produces the *operator Euler–Lagrange equations*

$$(q''; t'' |\delta S/\delta q(t)| q'; t')_s = 0 \qquad \text{for} \quad t' < t < t''. \qquad (2.25)$$

We obtain the Feynman operator equations

$$\int_{\mathcal{g}} [[\delta S/\delta q_i(t_1)]q_j(t_2) - i\hbar\, \delta_{ij}\, \delta(t_1 - t_2)](\exp iS/\hbar)D(\hat{q})$$

$$= (q''; t'' |[[\delta S/\delta q_i(t_1)]q_j(t_2) - i\hbar\, \delta_{ij}\, \delta(t_1 - t_2)]| q'; t')_s$$

$$= 0 \qquad \text{for} \quad t' < t_1, \quad t_2 < t'', \qquad (2.26)$$

by putting $F = q_j(t_2)(\exp iS/\hbar), j$ fixed, in Eq. (D.1). Such exact equations as (2.26) facilitate the determination of Feynman operators and the computation of terms in (2.22) without explicit functional integration. This is illustrated by example in Appendix F where we also derive the relationship between Feynman operators and the more conventional differential operators of quantum mechanics.

Feynman[9] has noted that it is also possible to formulate a "sum-over-histories" functional integral representation for the propagation kernel in terms of the canonical variables q, p. Instead of (2.7), the infinitesimal propagation kernel is postulated to have the form

$$K(q(t + \Delta t), q(t); \Delta t)$$

$$= \int \left(\exp i(p \cdot [q(t + \Delta t) - q(t)] - H([q(t + \Delta t) + q(t)]/2, p)\, \Delta t)/\hbar \right)$$

$$\times dp/(2\pi\hbar)^n, \qquad (2.27)$$

[9] R. P. Feynman, *Phys. Rev.* **84**, 108 (1951). Also see W. Tobocman, *Nuovo Cimento* 3, 1213 (1956).

where $H(q, p)$ is the Hamiltonian for the dynamical system and terms of the order $(\Delta t)^2$ are neglected. To distinguish the dummy integration variable in (2.27) for different values of t, it is convenient to put $p = p(t + \frac{1}{2} \Delta t)$. Then, by substituting the representation (2.27) for the infinitesimal propagation kernel into (2.6) and taking the limit $N \to \infty$, $\Delta t \to 0$, we obtain

$$K(q'', q'; t'' - t') = \int_{\mathscr{C} \times \mathscr{F}} (\exp i\bar{S}/\hbar)D(q, p), \qquad (2.28)$$

where the canonical variable action functional $\bar{S} = \bar{S}[q(t), p(t)]$ defined by (1.23) appears in the form of a Riemann sum,

$$\bar{S} = \lim_{\Delta t \to 0} \sum_{M=1}^{N = (t'' - t')/\Delta t} \left(p\left(t' + \left(M - \frac{1}{2}\right) \Delta t\right) \right.$$
$$\cdot [q(t' + M \Delta t) - q(t' + (M - 1) \Delta t)]$$
$$- H\left(\frac{q(t' + M \Delta t) + q(t' + (M - 1) \Delta t)}{2}, \right.$$
$$\left. p\left(t' + \left(M - \frac{1}{2}\right) \Delta t\right)\right) \Delta t\right), \qquad (2.29)$$

and $D(q, p)$ in (2.28) denotes an infinitesimal volume element for integration over the class of real n-tuple functions

$$\mathscr{C} = \{q = q(t) \quad \text{for} \quad t' \leqslant t \leqslant t'' : q(t') = q', q(t'') = q''\}$$

and the space of real n-tuple functions $\mathscr{F} = \{p = p(t) \text{ for } t' \leqslant t \leqslant t''\}$, the generalized momentum n-tuple being unconstrained at $t = t'$ and $t = t''$. In symbolic notation, we have

$$D(q, p) = \left(\prod_{t' < t < t''} dq(t) \right)\left(\prod_{t' \leqslant t \leqslant t''} [dp(t)/(2\pi\hbar)^n] \right) \qquad (2.30)$$

with the normalization given explicitly in the infinite product of momentum-space infinitesimal volume elements. For linear coordinate and momentum n-tuples, the q-space and p-space infinitesimal volume elements are $dq(t) = \prod_{i=1}^{n} dq_i(t)$ and $dp(t) = \prod_{i=1}^{n} dp_i(t)$, and the infinitesimal volume element (2.30) is *displacement-invariant* for any fixed $\sigma \in \mathscr{G} = \{\hat{q} = \hat{q}(t) \text{ for } t' \leqslant t \leqslant t'' : \hat{q}(t') = 0 = \hat{q}(t'')\}$ and $\omega \in \mathscr{F}$,

$$D(q + \sigma, p + \omega) = D(q, p). \qquad (2.31)$$

It is readily seen that the Feynman functional integral representation (2.28) yields the representation (2.13), with the normalization factor fixed completely, for dynamical systems described by a Hamiltonian of the form

$$H = \tfrac{1}{2} p \cdot p + a \cdot p + V \tag{2.32}$$

where $a = a(q)$ and $V = V(q)$; we have

$$\bar{S} = \int_{t'}^{t''} \left(p \cdot \dot{q} - \tfrac{1}{2} p \cdot p - a \cdot p - V \right) dt$$

$$= \int_{t'}^{t''} L \, dt - \int_{t'}^{t''} \tfrac{1}{2} (p + a - \dot{q}) \cdot (p + a - \dot{q}) \, dt \tag{2.33}$$

in which

$$L = L(q, \dot{q}) = \tfrac{1}{2} (\dot{q} - a) \cdot (\dot{q} - a) - V \tag{2.34}$$

is the Lagrangian associated with (2.32) by (1.22), and hence the functional integration over $\mathscr{F}$ in (2.28) simply produces

$$\int_{\mathscr{F}} \left(\exp - i \int_{t'}^{t''} \tfrac{1}{2} (p + a - \dot{q}) \cdot (p + a - \dot{q}) \, dt \right) \prod_{t' \le t \le t''} [dp(t)/(2\pi\hbar)^n]$$

$$\equiv \mathscr{N}, \tag{2.35}$$

the normalization factor in (2.10) depending on $(t'' - t')$ but being independent of $q(t)$. More generally, for a Hamiltonian which is not of the form (2.32), the dynamical laws (2.13) and (2.28) are inequivalent. Thus, for $H = (p \cdot p)^{1/2}$ with $n = 3$,(2.28) yields the propagation kernel[10]

$$K(q'', q'; t'' - t') = \lim_{\varepsilon \to 0} \frac{i(t'' - t')}{\pi^2 [(q'' - q') \cdot (q'' - q') - (t'' - t' - i\varepsilon)^2]^2}, \tag{2.36}$$

[10] The following symbolic evaluation of (2.28) with $n = 3$ and $H = (p \cdot p)^{1/2}$ can be justified by rigorous application of the definitions:

$$K(q'', q'; t'' - t')$$

$$= \int_{\mathscr{C} \times \mathscr{F}} \left(\exp i \int_{t'}^{t''} \frac{(p \cdot \dot{q} - (p \cdot p)^{1/2}) \, dt}{\hbar} \right) D(q, p)$$

$$= \int_{\mathscr{C}} \left(\prod_{t' \le t \le t''} \int \left(\exp i \frac{(p(t) \cdot \dot{q}(t) - (p(t) \cdot p(t))^{1/2}) \, dt}{\hbar} \right) \frac{dp(t)}{(2\pi\hbar)^3} \right) \prod_{t' < t < t''} dq(t)$$

$$= \lim_{\varepsilon \to 0} \int_{\mathscr{C}} \left(\prod_{t' \le t \le t''} \frac{i \, dt}{\pi^2 (\dot{q}(t) \cdot \dot{q}(t)(dt)^2 - (dt - i\varepsilon)^2)^2} \right) \prod_{t' < t < t''} dq(t)$$

$$= \lim_{\varepsilon \to 0} i \pi^{-2} (t'' - t') [(q'' - q') \cdot (q'' - q') - (t'' - t' - i\varepsilon)^2]^{-2},$$

while (2.13) is ill-defined for $H = (p \cdot p)^{1/2}$ with the associated Lagrangian vanishing identically. It is necessary to appeal to the observable physics in order to determine whether (2.13) or (2.28) is the appropriate dynamical law in the case of systems for which both Feynman functional integral representations exist and are unequal.

If the action evaluated at the classical "history" is large in absolute magnitude compared to $\hbar$, there is a quasiclassical approximation for the propagation kernel. To obtain this quasiclassical approximation, we evoke a Volterra expansion[11] of the action functional about $q_c = q_c(t)$,

$$S[q(t)] = S[q_c(t)] + \frac{1}{2} \int_{t'}^{t''} \int_{t'}^{t''} \tilde{q}(t_1) \cdot \frac{\delta^2 S}{\delta q(t_1)\, \delta q(t_2)}\Big|_{q=q_c}$$

$$\cdot \tilde{q}(t_2)\, dt_1\, dt_2 \qquad \text{plus} \quad \text{terms of higher order in } \tilde{q}(t),$$

$$(2.37)$$

in which $\tilde{q}(t) \equiv q(t) - q_c(t)$ with $q_c(t)$ satisfying (2.17) subject to the boundary conditions $q_c(t') = q'$ and $q_c(t'') = q''$. Thus Eq. (2.13), the Feynman functional integral representation with q a linear coordinate n-tuple, gives rise to the approximate expression

$$K(q'', q'; t'' - t') \cong (\exp iS[q_c(t)]/\hbar)$$

$$\times \int_{\mathscr{G}} \left(\exp i \int_{t'}^{t''} \int_{t'}^{t''} \tilde{q}(t_1) \cdot \frac{\delta^2 S}{\delta q(t_1)\, \delta q(t_2)}\Big|_{q=q_c} \cdot \tilde{q}(t_2)\, dt_1\, dt_2/2\hbar \right) D(\tilde{q}),$$

$$(2.38)$$

where in the spirit of the method of steepest descents (with $\hbar$ regarded as a small parameter) terms of higher order in $\tilde{q}(t)$ are dropped, and displacement-invariance of the Haar measure is used to replace $D(\hat{q})$

---

where the convolution integral relation

$$\lim_{\varepsilon \to 0} \int \frac{ia}{\pi^2[(\xi - \eta) \cdot (\xi - \eta) - (a - i\varepsilon)^2]^2} \frac{ib}{\pi^2[(\eta - \zeta) \cdot (\eta - \zeta) - (b - i\varepsilon)^2]^2}\, d\eta$$

$$= \lim_{\varepsilon \to 0} \frac{i(a + b)}{\pi^2[(\xi - \zeta) \cdot (\xi - \zeta) - (a + b - i\varepsilon)^2]^2}$$

holding for real 3-tuples ξ, η, ζ, and real positive a, b, is used.

[11] See Appendix A, Eq. (A.19).

by $D(\tilde{q})$. Except for an action functional which is a general quadratic expression in $q(t)$, the functional integral that remains in (2.38) depends on q' and q'' implicitly through a dependence on $q_c(t)$; the latter functional integral can be evaluated in the general case to yield the formal result[12]

$$K(q'', q'; t'' - t') \cong \mathcal{N}[\det(\delta^2 S_c)]^{-1/2}(\exp iS[q_c(t)]/\hbar), \qquad (2.39)$$

where $\mathcal{N} = \mathcal{N}(t'' - t')$ is a normalization factor and $\det(\delta^2 S_c)$ denotes the infinite product of all eigenvalues λ_μ of the symmetric matrix kernel $\delta^2 S[q(t)]/\delta q(t_1)\, \delta q(t_2)|_{q=q_c}$. A function of q', q'', and $(t'' - t')$, $\det(\delta^2 S_c) \equiv \prod_{\mu=1}^{\infty} \lambda_\mu$ must exist modulo normalization, that is, $\sum_{\mu=1}^{\infty} \ln(\alpha |\lambda_\mu|)$ must converge absolutely for some fixed $\alpha = \alpha(t'' - t')$, if the quasiclassical approximation (2.39) is to have applicability.

2. Schrödinger Formulation

Equation (2.13) is a general statement of the dynamical law in quantum mechanics for a system with a finite number of degrees of freedom. An alternative formulation of the dynamical law is obtained if (2.7) is used to derive a linear partial differential equation for the wave function rather than a functional integral representation for the propagation kernel. Amenable to a more systematic treatment with existing mathematical methods, integration of the dynamical equation for the wave function enables one to solve specific quantum mechanical problems without having to perform the general evaluation of (2.13).

By differentiating (2.1) with respect to t'' and letting $t' \to t''$ in the integral, we find that

$$\partial \psi(q''; t'')/\partial t'' = \int \Lambda(q'', q')\psi(q'; t'') \, dq' \qquad (2.40)$$

where

$$\Lambda(q'', q') \equiv \lim_{\Delta t \to 0} [\partial K(q'', q'; \Delta t)/\partial(\Delta t)]. \qquad (2.41)$$

---

[12] A similar evaluation is given by K. O. Friedrichs and H. N. Shapiro, "Seminar on Integration of Functionals," pp. I–19. Courant Institute of Mathematical Sciences, New York Univ., 1957.

To evaluate (2.41), we first express (2.7) in the form

$$K(q'', q'; \Delta t) = A^{-1} \exp(iL(q, \dot{q}) \,\Delta t / \hbar), \qquad (2.42)$$

where

$$A = A(\Delta t), \qquad q = (q'' + q')/2,$$

and

$$\dot{q} = (q'' - q')/\Delta t.$$

Since we have

$$\frac{\partial L(q, \dot{q})}{\partial (\Delta t)} = \frac{\partial \dot{q}}{\partial (\Delta t)} \cdot \frac{\partial L(q, \dot{q})}{\partial \dot{q}} = -\frac{(q'' - q')}{(\Delta t)^2} \cdot \frac{\partial L(q, \dot{q})}{\partial \dot{q}}$$

$$= -\frac{1}{(\Delta t)} \dot{q} \cdot \frac{\partial L(q, \dot{q})}{\partial \dot{q}}, \qquad (2.43)$$

it follows that

$$\frac{\partial K(q'', q'; \Delta t)}{\partial (\Delta t)} = \left[(i\hbar)^{-1} H - A^{-1} \frac{dA}{d(\Delta t)} \right] K(q'', q'; \Delta t), \quad (2.44)$$

where the Hamiltonian (1.18) appears in terms of q and $\dot{q}$,

$$H \equiv (\dot{q} \cdot \partial/\partial \dot{q} - 1) L(q, \dot{q}). \qquad (2.45)$$

In order to obtain (2.41) from (2.44) without mathematical ambiguity, it is necessary to eliminate $\dot{q} = (q'' - q')/\Delta t$, the quantity in (2.44) that is indeterminate in the limit $\Delta t \to 0$ with $\lim_{\Delta t \to 0} K(q'', q'; \Delta t) = \delta(q'' - q')$. To this end note that

$$\lim_{\Delta t \to 0} \frac{\partial K(q'', q'; \Delta t)}{\partial q''} = \frac{\partial \delta(q'' - q')}{\partial q''} \qquad (2.46)$$

can also be evaluated by using (2.42),

$$\lim_{\Delta t \to 0} \frac{\partial K(q'', q'; \Delta t)}{\partial q''} = \lim_{\Delta t \to 0} \left[\frac{i(\Delta t)}{\hbar} \frac{\partial L(q, \dot{q})}{\partial q''} K(q'', q'; \Delta t) \right]$$

$$= \frac{i}{\hbar} \left(\lim_{\Delta t \to 0} p \right) \delta(q'' - q'), \qquad (2.47)$$

where p is the generalized momentum n-tuple (1.17) expressed in terms of $q = (q'' + q')/2$ and $\dot{q} = (q'' - q')/\Delta t$. Relations (2.46) and (2.47) show that the generalized momentum n-tuple multiplied by $\delta(q'' - q')$ is well-defined as $\Delta t \to 0$ and given by

$$\left(\lim_{\Delta t \to 0} p\right) \delta(q'' - q') = -i\hbar[\partial \delta(q'' - q')/\partial q''].\qquad(2.48)$$

Thus, it would appear that the general procedure for securing (2.41) from (2.44) is to use (1.17) to eliminate $\dot{q}$ from (2.45), thereby obtaining $H = H(q, p)$, and then to replace p by the operator $-i\hbar\,\partial/\partial q''$ in the limit $\Delta t \to 0$, as prescribed by (2.48). However, since (2.42) also gives

$$\partial^2 K(q'', q'; \Delta t)/\partial q_i''^2 = \left[-\frac{1}{\hbar^2}\left(\frac{\partial L}{\partial \dot{q}_i}\right)^2 + \frac{i}{\hbar}\frac{\partial^2 L}{\partial q_i \partial \dot{q}_i} + \frac{i}{\hbar\,\Delta t}\frac{\partial^2 L}{\partial \dot{q}_i^2} + O(\Delta t)\right]$$

$$\times\, K(q'', q'; \Delta t),\qquad(2.49)$$

it is clear that in the limit $\Delta t \to 0$ powers of $p_i \equiv \partial L/\partial \dot{q}_i$ cannot be replaced simply by corresponding powers of $-i\hbar\,\partial/\partial q_i''$. It is the extra pure imaginary terms in the square bracket on the right-hand side of (2.49) which counter the normalization term in the square bracket on the right-hand side of (2.44) as $\Delta t \to 0$. Subject to a suitable renormalization transformation (2.15) [which induces $A \to (e^{-z\Delta t})A$ because of (2.11)], (2.41) becomes

$$\Lambda(q'', q') = \lim_{\Delta t \to 0}\left\{\left[(i\hbar)^{-1}H(q, p) - A^{-1}\frac{dA}{d(\Delta t)}\right]K(q'', q'; \Delta t)\right\}$$

$$= (i\hbar)^{-1}H\left(q'', -i\hbar\frac{\partial}{\partial q''}\right)\delta(q'' - q')\qquad(2.50)$$

for a certain symmetrical ordering of q, p product terms, if such terms with noncommuting factors occur in $H(q, p)$, so that $H(q, -i\hbar\,\partial/\partial q)$ is Hermitian on a space of sufficiently smooth absolute-square-integrable complex-valued functions of q.

To illuminate the preceding discussion and exemplify the cancellation effect in (2.50), let us consider the propagation kernel associated with a Lagrangian of the form (1.2), for which (2.42) becomes

$$K(q'', q'; \Delta t) = (2\pi i\hbar\,\Delta t)^{-n/2}$$

$$\times \exp\left(i\left[\frac{(q'' - q')\cdot(q'' - q')}{2\,\Delta t} - V(q)\,\Delta t\right]/\hbar\right)$$

$$(2.51)$$

with $q \equiv (q'' + q')/2$. For the Lagrangian (1.2) and $A = (2\pi i\hbar \, \Delta t)^{n/2}$, we have

$$(i\hbar)^{-1} H(q, p) - A^{-1}[dA/d(\Delta t)] = (i\hbar)^{-1}[\tfrac{1}{2} p \cdot p + V(q)] - n/2(\Delta t),$$

(2.52)

while (2.49) produces

$$\sum_{i=1}^{n} \partial^2 K(q'', q'; \Delta t)/\partial q_i''^2 = [- \hbar^{-2} p \cdot p + (in/\hbar \, \Delta t) + O(\Delta t)] K(q'', q'; \Delta t).$$

(2.53)

Thus, the terms in (2.52) yield a special case of (2.50),

$$\Lambda(q'', q') = (i\hbar)^{-1} \left[-\frac{\hbar^2}{2} \frac{\partial}{\partial q''} \cdot \frac{\partial}{\partial q''} + V(q'') \right] \delta(q'' - q'). \quad (2.54)$$

A similar cancellation eliminates the normalization term $A^{-1} \, dA/d(\Delta t)$ for any Lagrangian with (2.50) existing as a well-defined finite quantity. The previously undetermined complex constant in $\mathcal{N}$, associated with a renormalization transformation (2.15), if fixed by (2.50).

Now by putting (2.50) into (2.40), we obtain a linear partial differential dynamical equation for the wave function

$$i\hbar \, [\partial \psi(q; t)/\partial t] = H(q, -i\hbar \, \partial/\partial q) \psi(q; t), \quad (2.55)$$

the equation given by Schrödinger in 1925. The Feynman postulate for the infinitesimal propagation kernel (2.7) elucidates the origin and structure of the differential operator on the right-hand side of (2.55), the classical Hamiltonian suitably ordered and with p replaced by $-i\hbar \, \partial/\partial q$. The *iteration solution* for a wave function satisfying Eq. (2.55) with $H = H_0 + H_{int}$ is developed in Appendix H, along with the relationship of the iteration solution to the Feynman expansion (2.22) for the propagation kernel.

It should be noted that even without evoking a specific expression for the infinitesimal propagation kernel, the semigroup composition law (2.2) and the initial value condition (2.3) imply that the propagation kernel can be expressed formally in terms of an *exponentiated generator* as

$$K(q, q'; t - t') = (\exp -i(t - t')\mathbf{H}/\hbar) \, \delta(q - q'), \quad (2.56)$$

where the *quantum Hamiltonian* $\mathbf{H}$ is a linear differential operator in q and $\partial/\partial q$ that acts on the q in the δ-function. The quantum Hamiltonian $\mathbf{H}$ must be Hermitian for conservation of probability.[1] By putting (2.56) into (2.1), we obtain the formal dynamical equation for the wave function

$$\psi(q;t) = (\exp - i(t - t')\mathbf{H}/\hbar)\psi(q;t'),\qquad (2.57)$$

which gives the *abstract Schrödinger equation*

$$i\hbar[\partial\psi(q;t)/\partial t] = \mathbf{H}\psi(q;t)\qquad (2.58)$$

where the quantum Hamiltonian $\mathbf{H}$ still remains to be determined. It is the Feynman expression for the infinitesimal propagation kernel (2.7) that associates the quantum Hamiltonian with the classical Hamiltonian

$$\mathbf{H} = H(q, -i\hbar\,\partial/\partial q).\qquad (2.59)$$

With the state of the dynamical system represented by a wave function $\psi = \psi(q;t)$, the *expectation value* of an observable $f = f(q,p)$ is postulated as

$$\langle f\rangle \equiv \int \psi(q;t)^* f(q, -i\hbar\,\partial/\partial q)\psi(q;t)\,dq\qquad (2.60)$$

for a suitable (Hermitian and experimentally appropriate) ordering of q, p product terms, if such terms with noncommuting factors appear in $f(q,p)$, and where the wave function is assumed to be normalized to unity, $\int |\psi(q;t)|^2\,dq \equiv 1$. Again the Feynman formulation, and in particular Eq. (2.48), elucidates the origin and structure of the differential operator associated with an observable in (2.60).

Stationary states are described by wave functions of the form

$$\psi(q;t) = (\exp - iEt/\hbar)u(q)\qquad (2.61)$$

that satisfy (2.55), and hence $u(q)$ is an eigenfunction solution to the Schrödinger stationary state equation

$$\mathbf{H}u(q) \equiv H(q, -i\hbar\,\partial/\partial q)u(q) = Eu(q)\qquad (2.62)$$

associated with the constant energy eigenvalue E. If the quantum Hamiltonian in (2.62) admits a complete orthonormal set of complex-valued eigenfunctions $\{u_\mu(q)\}$ associated with the discrete set of energy eigenvalues $\{E_\mu\}$, we have

$$\mathbf{H}u_\mu(q) = E_\mu u_\mu(q),\qquad \int u_\mu{}^*(q)u_\nu(q)\,dq = \delta_{\mu\nu},$$
$$\sum_\mu u_\mu(q'')u_\mu{}^*(q') = \delta(q'' - q').\qquad (2.63)$$

Then it follows that the propagation kernel in (2.1) has the formal representation

$$K(q'', q'; t'' - t') = \sum_\mu (\exp - iE_\mu(t'' - t')/\hbar)u_\mu(q'')u_\mu{}^*(q').\quad (2.64)$$

Satisfying the initial value condition (2.3) because of the completeness relation in (2.63), the representation (2.64) is the formal solution to the propagation kernel equation implied by (2.55) and (2.1),

$$(i\hbar(\partial/\partial t'') - H(q'', -i\hbar\,\partial/\partial q''))K(q'', q'; t'' - t') = 0 \qquad (t'' > t'). \quad (2.65)$$

In general, a quantum Hamiltonian operator has a spectrum of energy eigenvalues that is bounded from below, an obvious requisite for physically realizable dynamics. With only one eigenfunction associated with the *ground state energy* $E_0 \equiv \min_\mu \{E_\mu\}$, (2.64) yields the formulas

$$E_0 = -\lim_{s\to\infty} (\hbar/s) \ln K(q'', q'; -is), \qquad (2.66)$$

$$u_0(q'')u_0^*(q') = \lim_{s\to\infty} [(\exp E_0 s/\hbar)K(q'', q'; -is)]. \qquad (2.67)$$

Similar relations for the E_μ and $u_\mu(q)$ of other stationary states are obtained by successively removing terms from (2.64) and rewriting (2.66) and (2.67) with the subtracted propagation kernels. It is a straightforward matter to generalize these formulas to handle *degenerate energy multiplets*, that is, several linearly independent $u_\mu(q)$ associated with the same energy eigenvalue. Thus, energy eigenvalues and eigenfunctions can be extracted by simple limiting procedures from a closed-form expression for the propagation kernel, as provided by the Feynman functional integral representation (2.13). For example, from the propagation kernel (2.19) with (2.21), we obtain

$$E_0 = -\lim_{s\to\infty} \frac{\hbar}{s} \sum_{i=1}^{n} \left[\ln\left(\frac{\omega_i}{2\pi\hbar \sinh \omega_i s}\right)^{1/2} \right.$$
$$\left. - \frac{\omega_i}{\hbar}\left(\frac{q_i''^2 + q_i'^2}{2\tanh \omega_i s} - \frac{q_i''q_i'}{\sinh \omega_i s}\right)\right] = \frac{1}{2}\hbar\sum_{i=1}^{n}\omega_i, \quad (2.68)$$

$$u_0(q'')u_0^*(q') = \lim_{s\to\infty} \left[\prod_{i=1}^{n}\left(\exp\frac{1}{2}\omega_i s\right)\left(\frac{\omega_i}{2\pi\hbar \sinh \omega_i s}\right)^{1/2} \right.$$
$$\left. \times \left(\exp -\frac{\omega_i}{\hbar}\left(\frac{q_i''^2 + q_i'^2}{2\tanh \omega_i s} - \frac{q_i''q_i'}{\sinh \omega_i s}\right)\right)\right]$$
$$= \prod_{i=1}^{n}\left(\frac{\omega_i}{\pi\hbar}\right)^{1/2}\left(\exp -\frac{\omega_i}{2\hbar}(q_i''^2 + q_i'^2)\right) \quad (2.69)$$

$$\Rightarrow \qquad u_0(q) = \text{(phase const)} \prod_{i=1}^{n}\left(\frac{\omega_i}{\pi\hbar}\right)^{1/4}\left(\exp -\frac{\omega_i}{2\hbar}q_i^2\right). \quad (2.70)$$

The Schrödinger formulation also supplies a dynamical description for systems without a classical analog, systems that cannot be treated directly by the Feynman formulation with the propagation kernel given by (2.13) or (2.28). Equations (2.56), (2.57), and (2.58) hold good for such systems, but the quantum Hamiltonian **H** must be prescribed without appeal to an equation of the form (2.59). We have

$$\mathbf{H} = -i\hbar \begin{pmatrix} \partial/\partial q_3 & \partial/\partial q_1 - i\,\partial/\partial q_2 \\ \partial/\partial q_1 + i\,\partial/\partial q_2 & -\partial/\partial q_3 \end{pmatrix} \qquad (2.71)$$

in the case of Pauli's neutrino with $\psi(q;t)$ a two-component wave function, and

$$\mathbf{H} = -i\hbar \begin{pmatrix} 1 & 0 \\ 0 & -1 \end{pmatrix} \times \begin{pmatrix} \partial/\partial q_3 & \partial/\partial q_1 - i\,\partial/\partial q_2 \\ \partial/\partial q_1 + i\,\partial/\partial q_2 & -\partial/\partial q_3 \end{pmatrix}$$
$$+ \mu \begin{pmatrix} 0 & 1 \\ 1 & 0 \end{pmatrix} \times \begin{pmatrix} 1 & 0 \\ 0 & 1 \end{pmatrix} \qquad (2.72)$$

in the case of Dirac's electron with $\psi(q;t)$ a four-component wave function. The propagation kernel (2.56) associated with either (2.71) or (2.72) assumes the generic form

$$K(q'', q'; t'' - t') = \int (\exp i(k \cdot (q'' - q') - (t'' - t')\hat{H}(\hbar k)/\hbar)) \, [dk/(2\pi)^3], \qquad (2.73)$$

in which $\hat{H}(\hbar k)$ is the Hermitian matrix obtained by letting **H** act on $(\exp ik \cdot q)$: $\mathbf{H}(\exp ik \cdot q) = \hat{H}(\hbar k)(\exp ik \cdot q)$. Equation (2.4) retains validity for such a matrix-valued propagation kernel if the infinitesimal propagation kernels are appropriately ordered by setting $K(q^{(M+1)}, q^{(M)}; \Delta t)$ to the left of $K(q^{(M)}, q^{(M-1)}; \Delta t)$ for each value of M successively. By putting $\hbar k \equiv p$ in (2.73), we can express the infinitesimal propagation kernel as

$$K(q^{(M)}, q^{(M-1)}; \Delta t)$$
$$= \int (\exp i(p \cdot (q^{(M)} - q^{(M-1)}) - \hat{H}(p)\,\Delta t)/\hbar) \, [dp/(2\pi\hbar)^3], \qquad (2.74)$$

a representation similar in form to (2.27). It follows that the propagation kernel can be prescribed as a functional integral with the integrand ordered chronologically,

$K(q'', q'; t'' - t')$

$$= \int_{\mathscr{C} \times \mathscr{F}} \left(T \exp i \int_{t'}^{t''} (p(t) \cdot \dot{q}(t) - \hat{H}(p(t))) \, dt/\hbar \right) D(q, p).$$

$$(2.75)$$

Here, the chronological ordering symbol T (defined as in Eq. (H.7), Appendix H) arranges noncommuting factors (that is, $\hat{H}(p(t))$ for distinct t) to the left with increasing values of t, and the other notation in (2.75) is the same as in (2.28). In a generalized sense, the functional integral (2.75) is a "sum-over-histories" representation for the propagation kernel[13] in terms of the "canonical variables" q, p. Assuming that we have a linear momentum n-tuple and a matrix-valued $\hat{H}(p)$ of the form associated with (2.71) or (2.72), it is possible to perform the functional integrations in (2.75) and arrive at an explicit closed-form expression for the propagation kernel analogous to (2.36). However, the chronological ordering of the integrand and matrix quality of $\hat{H}(p(t))$ in (2.75) creates an additional technical difficulty which cohibits evaluation of the functional integral for more general quantum Hamiltonians.

3. Dirac Formulation

By evoking the dynamical equation (2.57), the generic expectation value equation (2.60) can be recast in the form

$$\langle f \rangle = \int \psi(q; 0)^* \mathbf{f} \psi(q; 0) \, dq, \qquad (2.76)$$

with $\mathbf{H}$ defined by (2.59) and the Hermitian *observable*

$$\mathbf{f} \equiv (\exp it\mathbf{H}/\hbar)f(q, -i\hbar \, \partial/\partial q)(\exp -it\mathbf{H}/\hbar) = f(\mathbf{q}, \mathbf{p}), \qquad (2.77)$$

$$\mathbf{q} \equiv (\exp it\mathbf{H}/\hbar)q(\exp -it\mathbf{H}/\hbar) \qquad (2.78)$$

$$\mathbf{p} \equiv (\exp it\mathbf{H}/\hbar)(-i\hbar \, \partial/\partial q)(\exp -it\mathbf{H}/\hbar), \qquad (2.79)$$

embodying the quantum mechanical dynamics, and the wave function in (2.76) fixed at the initial instant of time $t = 0$. Equations (2.76) and (2.77) are mathematical statements for the *Heisenberg picture* of quantum

---

[13] The academic case of one spatial dimension admits a more direct representation for the propagation kernel associated with the Dirac equation, a "relativistic sum over histories" for the one-dimensional Dirac equation described in Appendix E.

mechanics. The value of the Heisenberg picture is that the time rate of change of any expectation value (2.76),

$$\langle \hat{f} \rangle = \int \psi(q; 0)^* \mathbf{f} \psi(q; 0) \, dq, \tag{2.80}$$

can be obtained by differentiating (2.77),

$$df/dt \equiv \dot{\mathbf{f}} = [\mathbf{f}, \mathbf{H}], \tag{2.81}$$

with the *Dirac bracket* of two observable operators introduced as

$$[\mathbf{f}, \mathbf{g}] \equiv (i\hbar)^{-1}(\mathbf{fg} - \mathbf{gf}). \tag{2.82}$$

Thus, for example, the components of (2.78) and (2.79) have the Dirac brackets

$$[\mathbf{q}_i, \mathbf{q}_j] = [\mathbf{p}_i, \mathbf{p}_j] = 0, \qquad [\mathbf{q}_i, \mathbf{p}_j] = \delta_{ij}, \tag{2.83}$$

with the identity operator understood to appear on the right-hand side of the latter equation. The observable constants of the motion are quantities which have a zero Dirac bracket with the quantum Hamiltonian (2.59). It should be noted that no ambiguity arises in our using square-bracket notation for both the Poisson and Dirac combination laws (1.26) and (2.82) since the former is understood to apply in the case of classical observables, while the latter is understood to apply in the case of quantum observables (boldface type).

It is easy to verify that the Dirac bracket combination law (2.82), associating a new Hermitian observable operator with an ordered pair of Hermitian observable operators, has all the properties required of a combination law for a Lie algebra, properties (1.28)–(1.30). Furthermore, the set of all Hermitian observable operators is closed under operator addition and Dirac bracket combination, and hence constitutes a suitable set for a Lie algebra with the Dirac bracket combination law. Equations (1.25) and (2.81) for the time rate of change of observables are identical except for the meaning of the bracket combination law, according to Poisson in classical mechanics with (1.26) and according to Dirac in quantum mechanics with (2.82). A Lie algebraic structure for the set of all observables is provided in either case by the bracket combination law.[14] However, the entire Lie algebraic structure in

[14] The abstract Lie algebraic relationship between classical mechanics and quantum mechanics was discussed by T. F. Jordan and E. C. G. Sudarshan, *Rev. Modern Phys.* **33**, 515 (1961). More concrete formalisms which embrace both fundamental dynamical theories were presented subsequently by G. Rosen, *J. Franklin Inst.* **279**, 457 (1965); F. Strocchi, *Rev. Modern Phys.* **38**, 36 (1966). The author's formalism for classical and quantum statistical mechanics is described in Appendix G.

classical mechanics with (1.26) cannot be preserved in quantum mechanics with (2.82) by the *Dirac correspondence*

$$[f, g] = h \Rightarrow [\mathbf{f}, \mathbf{g}] = \mathbf{h} \qquad (2.84)$$

even though extra special properties of the Poisson bracket, such as (1.31), hold good with proper ordering under the Dirac correspondence. In general, only a certain elite subset of the observables follows the Dirac correspondence (2.84), in the sense that in the subset any triplet of observables f, g, h related by the Poisson bracket yields a triplet $\mathbf{f}, \mathbf{g}, \mathbf{h}$ related by the Dirac bracket.[15] It is important to note that observables of the form (1.27), elements of the $(2n^2 + 3n + 1)$-dimensional Lie algebra $\mathscr{A}_s$, follow the Dirac correspondence (2.84). This is verified most readily by using the phase space notation of Eqs. (1.46) and (1.47) to express an element of $\mathscr{A}_s$ as

$$f = \tfrac{1}{2} f_{ab}^{(2)} X_a X_b + f_a^{(1)} X_a + f^{(0)}, \qquad (2.85)$$

where repeated indices are understood to be summed 1 to $2n$, and the $2n \times 2n$ array of real constants $f_{ab}^{(2)} \equiv f_{ba}^{(2)}$ is symmetrical; the Poisson bracket (1.47) associated with two observables f and g of the form (2.85) is

$$[f, g] = \tfrac{1}{2}(f_{ac}^{(2)}\Omega_{cd} g_{db}^{(2)} + f_{bc}^{(2)}\Omega_{cd} g_{da}^{(2)})X_a X_b + (f_b^{(1)}\Omega_{bc} g_{ca}^{(2)}$$
$$+ f_{ab}^{(2)}\Omega_{bc} g_c^{(1)})X_a + f_a^{(1)}\Omega_{ab} g_b^{(1)}$$

$$\equiv h. \qquad (2.86)$$

---

[15] R. Arens and D. Babbitt, *J. Math. Phys.* 6, 1071 (1965) and works cited therein. We are *not* concerned here with the academic "Dirac problem" (see, for example, R. F. Streater, *Comm. Math. Phys.* 2, 354 (1966) and works cited therein): To find a mapping $f \to \mathbf{f}$, an Hermitian differential operator assignment other than (2.77), for which the entire Lie algebraic structures in classical mechanics and quantum mechanics are isomorphic. One solution to the academic "Dirac problem" is given by the mapping

$$\mathbf{f} \equiv f - \frac{1}{2} X_a \frac{\partial f}{\partial X_a} + i\hbar \frac{\partial f}{\partial X_a} \Omega_{ab} \frac{\partial}{\partial X_b},$$

a differential operator in the phase space variables that is Hermitian with the Poincaré invariant inner product definition; by calculating the Dirac bracket (2.82) for two such Hermitian "observable" operators, we find

$$[\mathbf{f}, \mathbf{g}] = [f, g] - \frac{1}{2} X_a \frac{\partial [f, g]}{\partial X_a} + i\hbar \frac{\partial [f, g]}{\partial X_a} \Omega_{ab} \frac{\partial}{\partial X_b}$$

where the Poisson bracket appears as (1.47), and hence (2.84) is satisfied by this mapping for all "observables."

According to (2.77), the Hermitian operator

$$\mathbf{f} = \tfrac{1}{2}f_{ab}^{(2)}\mathbf{X}_a\mathbf{X}_b + f_a^{(1)}\mathbf{X}_a + f^{(0)} \tag{2.87}$$

follows from (2.85); the Dirac bracket (2.82) associated with two observables $\mathbf{f}$ and $\mathbf{g}$ of the form (2.87) is

$$[\mathbf{f}, \mathbf{g}] = \tfrac{1}{2}(f_{ac}^{(2)}\Omega_{cd}\,g_{db}^{(2)} + f_{bc}^{(2)}\Omega_{cd}\,g_{da}^{(2)})\mathbf{X}_a\mathbf{X}_b$$
$$+ (f_b^{(1)}\Omega_{bc}\,g_{ca}^{(2)} + f_{ab}^{(2)}\Omega_{bc}\,g_c^{(1)})\mathbf{X}_a + f_a^{(1)}\Omega_{ab}\,g_b^{(1)} = \mathbf{h}, \tag{2.88}$$

where the final member is the Hermitian operator associated with (2.86). In obtaining (2.88), we employ the Dirac bracket relations

$$[\mathbf{X}_a, \mathbf{X}_b] = \Omega_{ab} \tag{2.89}$$

equivalent to Eqs. (2.83). It follows from (2.88) that the subset of Hermitian observables having the generic form (2.87) is closed with respect to Dirac bracket combination, and hence constitutes a $(2n^2 + 3n + 1)$-dimensional Lie algebra $\mathscr{A}_s$, the quantum correspondent of $\mathscr{A}_s$. Moreover, a comparison of (2.86) and (2.88) shows that the Lie algebraic structure of $\mathscr{A}_s$ with the Poisson bracket combination law is identical to the Lie algebraic structure of $\mathscr{A}_s$ with the Dirac bracket combination law: The Lie algebras $\mathscr{A}_s$ and $\mathscr{A}_s$ are isomorphic. However, it is apparent that observables with higher order q, p product terms cannot be assigned an Hermitian ordering which validates the Dirac correspondence (2.84) for them and other observables. For example, consider the observable

$$f = q^2p^2 \qquad \text{with} \quad n = 1, \tag{2.90}$$

given by the Poisson bracket

$$f = \tfrac{1}{9}[q^3, p^3] \tag{2.91}$$

and satisfying the double Poisson bracket equation

$$f = \tfrac{1}{24}[[q^2, f], p^2]. \tag{2.92}$$

If we apply the Dirac correspondence (2.84) to (2.91) and evaluate the Dirac bracket by using (2.83) with $i = j = 1$, we find

$$\mathbf{f} = \tfrac{1}{3}(\mathbf{qpqp} + \mathbf{pqpq}) + \tfrac{1}{6}(\mathbf{q}^2\mathbf{p}^2 + \mathbf{p}^2\mathbf{q}^2)$$
$$= \tfrac{1}{2}(\mathbf{q}^2\mathbf{p}^2 + \mathbf{p}^2\mathbf{q}^2) + \tfrac{1}{3}\hbar^2, \tag{2.93}$$

while on the other hand, by applying the Dirac correspondence to (2.92), we have

$$\mathbf{f} = \tfrac{1}{24}[[\mathbf{q}^2, \mathbf{f}], \mathbf{p}^2], \tag{2.94}$$

which implies that

$$\mathbf{f} = \tfrac{1}{2}(\mathbf{qpqp} + \mathbf{pqpq}) = \tfrac{1}{2}(\mathbf{q}^2\mathbf{p}^2 + \mathbf{p}^2\mathbf{q}^2) + \tfrac{1}{2}\hbar^2, \tag{2.95}$$

an Hermitian ordering for $\mathbf{f}$ that differs from (2.93). Hence, it is impossible to prescribe an Hermitian ordering for the observable $\mathbf{f}$ associated with (2.90) that validates the Dirac correspondence (2.84) for all $\mathbf{g}$. This situation prevails generally for observables with $\mathbf{q}, \mathbf{p}$ product terms of higher order than bilinear in the canonical variables.

To the Lie algebra $\mathscr{A}$ of a subset of Hermitian observables that is closed with respect to operator addition and Dirac bracket combination, there is associated a conjugate Lie algebra $\mathscr{A}^*$ consisting of all observables which have a zero Dirac bracket with all elements of $\mathscr{A}$. Such a Lie algebra $\mathscr{A}^*$ is infinite-dimensional because arbitrary powers and symmetrized products of Hermitian observables in $\mathscr{A}^*$ are also contained in $\mathscr{A}^*$. The observable constants of the motion constitute the Lie algebra $\mathscr{A}_H^* \equiv \{\mathbf{f}: [\mathbf{f}, \mathbf{H}] = 0\}$ conjugate to the 1-dimensional Lie algebra consisting of elements proportional to the quantum Hamiltonian, $\mathscr{A}_H \equiv \{(\text{real constant parameter}) \times \mathbf{H}\}$. Associated with the infinite-dimensional Lie algebra $\mathscr{A}_H^*$, we have the finite-dimensional Lie algebra

$$\hat{\mathscr{A}}_{\mathbf{H}}^* \equiv \left\{ \left(\sum_{i=1}^{m} \alpha_i w_i \right) : [\mathbf{w}_i, \mathbf{H}] = 0 \right\},$$

where the α's are real constants, and the $\mathbf{w}$'s (with no $\mathbf{w}_i$ merely a function of $\mathbf{H}$) constitute a maximal finite set, closed under Dirac bracket combination, of linearly independent constants of the motion (exclusive of $\mathbf{H}$). In general, the classical and quantum Lie algebras $\mathscr{A}_H^*$ and $\mathscr{A}_H^*$ are not isomorphic for there occur elements of $\mathscr{A}_H^*$ with q, p product terms of higher order than bilinear in the canonical variables.[16] However, for dynamical systems with the ordinary types of physical symmetry, it is usually possible to prescribe Hermitian orderings which take each $w_i = w_i(q, p)$ of $\hat{\mathscr{A}}_H^*$ into a $\mathbf{w}_i = w_i(\mathbf{q}, \mathbf{p})$ of $\hat{\mathscr{A}}_H^*$

---

[16] Necessary conditions for establishing isomorphism of the Lie algebras $\mathscr{A}_H^*$ and $\mathscr{A}_H^*$ have been discussed by H. Narumi, *J. Phys. Soc. Japan* **11**, 786, 1956.

in such a manner that $\mathscr{A}_H{}^*$ and $\hat{\mathscr{A}}_H{}^*$ are isomorphic, the Dirac correspondence (2.84) holding for all associated observables in the Lie algebras. The usual isomorphism of $\mathscr{A}_H{}^*$ and $\hat{\mathscr{A}}_H{}^*$ is illustrated by the n-dimensional hydrogen atom Hamilton, the *generalized Kepler Hamiltonian*[17]

$$H = \tfrac{1}{2}p \cdot p - k(q \cdot q)^{-1/2}. \qquad (2.96)$$

There are $\frac{1}{2}(n^2 + n)$ w's in the maximal finite set of independent constants of the motion generated by (2.96), namely the generalized Runge–Lenz observables[18]

$$w_i = (-2H)^{-1/2}[(q \cdot p)p_i - (p \cdot p)q_i + k(q \cdot q)^{-1/2}q_i] \qquad (2.97)$$

in addition to the $\frac{1}{2}(n^2 - n)$ generalized angular momentum observables of the form $(q_i p_j - q_j p_i)$; the set of all real linear combinations of the $\frac{1}{2}(n^2 + n)$ w's is closed under Poisson bracket combination and constitutes the Lie algebra $\mathscr{A}_H{}^*$, which can be shown to be isomorphic to the Lie algebra associated with the group of all $(n + 1) \times (n + 1)$ real orthogonal matrices of determinant unity, the Lie group $SO(n + 1)$. The orderings[19]

$$\mathbf{w}_i = (-2\mathbf{H})^{-1/2}\{\tfrac{1}{2}(\mathbf{q} \cdot \mathbf{p})p_i + \tfrac{1}{2}p_i(\mathbf{p} \cdot \mathbf{q})$$
$$- \tfrac{1}{2}(\mathbf{p} \cdot \mathbf{p})\mathbf{q}_i - \tfrac{1}{2}\mathbf{q}_i(\mathbf{p} \cdot \mathbf{p}) + k(\mathbf{q} \cdot \mathbf{q})^{-1/2}\mathbf{q}_i\} \qquad (2.98)$$

are Hermitian because $\mathbf{H}$ commutes with the curly-bracketed quantity in (2.98), and the observables (2.98) together with the observables $(\mathbf{q}_i \mathbf{p}_j - \mathbf{q}_j \mathbf{p}_i)$ yield a $\mathscr{A}_H{}^*$ isomorphic to $\hat{\mathscr{A}}_H{}^*$, hence, to the Lie algebra for $SO(n + 1)$.

A transformation of all observables

$$\mathbf{f} = f(\mathbf{q}, \mathbf{p}) \overset{\mathbf{w}}{\to} \mathbf{f}_\tau = f_\tau(\mathbf{q}, \mathbf{p})$$
$$\mathbf{g} = g(\mathbf{q}, \mathbf{p}) \overset{\mathbf{w}}{\to} \mathbf{g}_\tau = g_\tau(\mathbf{q}, \mathbf{p}) \qquad (2.99)$$
$$\cdots$$

such that

$$\mathbf{f}_0 = \mathbf{f}, \qquad \mathbf{g}_0 = \mathbf{g}, \qquad \cdots \qquad (2.100)$$

---

[17] See Footnote 5, Chapter 1.

[18] In order for the quantities (2.97) to be real observables, the canonical energy must be negative; then (2.97) applies to the phase space region for which $H < 0$. In the case of a positive canonical energy, the Runge–Lenz observables are defined with the prefactor $(2H)^{-1/2}$.

[19] This generalizes the $n = 3$ case discussed by V. Fock, *Z. Physik* **98**, 145, 1935; V. Bargmann, *Z. Physik* **99**, 576, 1936.

is said to be a *canonical transformation* generated by the Hermitian observable $\mathbf{w} = w(q, -i\hbar\, \partial/\partial q)$, considered here to be independent of t, if the transformed observables satisfy the first-order total differential equations

$$d\mathbf{f}_\tau/d\tau = [\mathbf{f}_\tau, \mathbf{w}], \qquad d\mathbf{g}_\tau/d\tau = [\mathbf{g}_\tau, \mathbf{w}], \qquad \cdots \qquad (2.101)$$

A transformed observable is given explicitly in terms of the original observable by the Maclaurin series derived from (2.100) and (2.101),

$$\mathbf{f}_\tau = \mathbf{f} + \tau[\mathbf{f}, \mathbf{w}] + (\tau^2/2!)[[\mathbf{f}, \mathbf{w}], \mathbf{w}] + \cdots = (\exp i\tau\mathbf{w}/\hbar)\mathbf{f}(\exp -i\tau\mathbf{w}/\hbar).$$
$$(2.102)$$

In analogy to the preservation of Poisson bracket relations by a classical canonical transformation, for a quantum canonical transformation we have the preservation of Dirac bracket relations,

$$\mathbf{h} = [\mathbf{f}, \mathbf{g}] \Rightarrow \mathbf{h}_\tau = [\mathbf{f}_\tau, \mathbf{g}_\tau], \qquad (2.103)$$

as an immediate consequence of (2.102). Thus, canonical transformations preserve the Lie algebraic structure for a subset of observables that is closed with respect to addition and Dirac bracket combination: Lie algebras are mapped into isomorphic Lie algebras by canonical transformations. Again in analogy to classical dynamics, the time rate of change of any canonically transformed observable is prescribed by a dynamical equation of the form (2.81),

$$d\mathbf{f}_\tau/dt \equiv \dot{\mathbf{f}}_\tau = [\mathbf{f}_\tau, \mathbf{H}_\tau], \qquad (2.104)$$

as seen by putting $\mathbf{h} = \dot{\mathbf{f}}$ and $\mathbf{g} = \mathbf{H}$ in (2.103), or by recalling (2.77) to express (2.102) as

$$\mathbf{f}_\tau = (\exp i\tau\mathbf{w}/\hbar)(\exp it\mathbf{H}/\hbar)f(q, -i\hbar\, \partial/\partial q)(\exp -it\mathbf{H}/\hbar)(\exp -i\tau\mathbf{w}/\hbar)$$

$$= (\exp it\mathbf{H}_\tau/\hbar)(\exp i\tau\mathbf{w}/\hbar)f(q, -i\hbar\, \partial/\partial q)(\exp -i\tau\mathbf{w}/\hbar)(\exp -it\mathbf{H}_\tau/\hbar)$$
$$(2.105)$$

where

$$\mathbf{H}_\tau \equiv (\exp i\tau\mathbf{w}/\hbar)\mathbf{H}(\exp -i\tau\mathbf{w}/\hbar). \qquad (2.106)$$

Next, by specializing (2.103) to the cases for which $\mathbf{h} = c$, a numerical constant (that is, a constant multiple of the identity operator), we note that (2.102) produces $\mathbf{h}_\tau = c$ for $\tau \geq 0$, and hence

$$[\mathbf{f}, \mathbf{g}] = c \Rightarrow [\mathbf{f}_\tau, \mathbf{g}_\tau] = c. \qquad (2.107)$$

In particular, for the components of the generalized coordinate n-tuple and generalized momentum n-tuple, the Dirac bracket equations (2.83) yield

$$[q_{\tau i}, q_{\tau j}] = [p_{\tau i}, p_{\tau j}] = 0, \qquad [q_{\tau i}, p_{\tau j}] = \delta_{ij}. \qquad (2.108)$$

The quantum analog of Eq. (1.45), which states that the Poisson bracket of two observables with respect to q_τ and p_τ equals their Poisson bracket computed with respect to q and p, is simply manifest in the fact that the Dirac bracket definition does not involve the canonical variables explicitly. Specializing to the cases for which f_τ equals each of the $2n$ components of q_τ and p_τ in (2.104), we immediately obtain dynamical equations for the transformed canonical variables,

$$\dot{q}_\tau = [q_\tau, H_\tau], \qquad \dot{p}_\tau = [p_\tau, H_\tau]. \qquad (2.109)$$

To make canonical transformations especially useful in quantum mechanics, one defines the transformed wave function at the initial instant of time as

$$\psi_\tau(q; 0) \equiv (\exp i\tau w/\hbar)\psi(q; 0), \qquad (2.110)$$

so that expectation values (2.76) are invariant with respect to canonical transformations,

$$\int \psi_\tau(q; 0)^* f_\tau \psi_\tau(q; 0) \, dq = \langle f \rangle. \qquad (2.111)$$

Hence, canonical transformations in quantum mechanics have no effect on observable predictions of the theory.

As in classical mechanics, the notion of canonical transformations generated by an observable can be generalized to the notion of canonical transformations generated by a finite-dimensional Lie algebra. Consider an m-dimensional Lie algebra $\mathscr{A}$ composed of all real linear combinations of m linearly independent Hermitian observables $w_1, \ldots, w_m$ with each $w_i = w_i(q, -i\hbar \, \partial/\partial q)$ independent of time. Since $\mathscr{A}$ is closed with respect to Dirac bracket combination, we must have

$$[w_i, w_j] = -\sum_{k=1}^{m} c_{ijk} w_k, \qquad (2.112)$$

where the $c_{ijk} = -c_{jik}$ are structure constants of the Lie algebra satisfying the quadratic Lie identities (1.52). We introduce the generators for quantum canonical transformations as

$$G_i \equiv (i/\hbar)w_i = (i/\hbar)w_i(q, -i\hbar \, \partial/\partial q), \qquad (2.113)$$

the m partial differential operators that satisfy the commutation relations

$$\mathbf{G}_i \, \mathbf{G}_j - \mathbf{G}_j \, \mathbf{G}_i = \sum_{k=1}^{m} c_{ijk} \, \mathbf{G}_k \tag{2.114}$$

as a consequence of (2.112) and the Dirac bracket definition (2.82). The classical generators (1.53) differ strikingly from the quantum generators (2.113) since the latter exhibit a general nonlinear dependence on $\partial/\partial q$ with p and $\partial/\partial p$ absent. A linear representation of the m-parameter Lie group associated with the Lie algebra $\mathscr{A}$ is provided by linear differential operators of the form

$$(\exp \alpha \cdot \mathbf{G}) \equiv \sum_{N=0}^{\infty} (N!)^{-1} (\alpha \cdot \mathbf{G})^N,$$

parametrized by a real m-tuple $\alpha = (\alpha_1, \ldots, \alpha_m)$. Under canonical transformations generated by $\mathscr{A}$, a generic Hermitian observable (2.77) is transformed to

$$\mathbf{f}_\alpha = (\exp \alpha \cdot \mathbf{G}) \mathbf{f} (\exp -\alpha \cdot \mathbf{G}), \tag{2.115}$$

as seen by putting $\tau = 1$ and $\mathbf{w} = \sum_{i=1}^{m} \alpha_i \mathbf{w}_i$ in (2.102). Conversely, by setting $\alpha = \tau \beta$ and $\sum_{i=1}^{m} \beta_i \mathbf{w}_i = \mathbf{w}$ with β a fixed unit m-tuple, we recover (2.102) from (2.115). It follows that the m-parameter canonical transformations generated by $\mathscr{A}$ produce transformed Hermitian observables (2.115) with all the properties derived for the transformed observables (2.102); the m-parameter canonical transformation generalization of (2.103) is

$$\mathbf{h} = [\mathbf{f}, \mathbf{g}] \Rightarrow \mathbf{h}_\alpha = [\mathbf{f}_\alpha, \mathbf{g}_\alpha], \tag{2.116}$$

the implied generalization of (2.104) is

$$d\mathbf{f}_\alpha/dt \equiv \dot{\mathbf{f}}_\alpha = [\mathbf{f}_\alpha, \mathbf{H}_\alpha], \tag{2.117}$$

and the transformed wave function follows from (2.110) and (2.113) as

$$\psi_\alpha(q; 0) = (\exp \alpha \cdot \mathbf{G}) \psi(q; 0). \tag{2.118}$$

An illustrative example of quantum canonical transformations generated by a finite-dimensional Lie algebra is provided by the quantum correspondent of (1.58),

$$\mathscr{A} = \left\{ \alpha_1 \mathbf{w}_1 + \alpha_2 \mathbf{w}_2 + \alpha_3 \mathbf{w}_3 \colon \mathbf{w}_1 \equiv -\frac{1}{4} \left(\hbar^2 \frac{\partial}{\partial q} \cdot \frac{\partial}{\partial q} + q \cdot q \right), \right.$$

$$\left. \mathbf{w}_2 \equiv \frac{1}{4} \left(\hbar^2 \frac{\partial}{\partial q} \cdot \frac{\partial}{\partial q} - q \cdot q \right), \quad \mathbf{w}_3 \equiv \frac{1}{4} i\hbar \left(q \cdot \frac{\partial}{\partial q} + \frac{\partial}{\partial q} \cdot q \right) \right\}.$$

$$\tag{2.119}$$

For this $\mathcal{A}$ the Dirac bracket closure relations (2.112) are

$$[w_1, w_2] = -w_3, \qquad [w_2, w_3] = -w_1,$$
$$[w_3, w_1] = w_2,$$

(2.120)

and thus the independent nonzero structure constants are

$$c_{123} = c_{231} = -c_{312} = 1.$$

Because the Lie algebra (2.119) is a subalgebra of $\mathcal{A}_s$, it is isomorphic to the corresponding subalgebra of $\mathcal{A}_s$, namely (1.58), and thus $\mathcal{A}$ defined by (2.119) is isomorphic to the Lie algebra for $SL(2, R)$. Generators of the form (2.113) for this $\mathcal{A}$ give a representation of $SL(2, R)$ canonical transformations on the space of wave functions, as prescribed by (2.118).

Special importance is attached to canonical transformations generated by the quantum Hamiltonian and by the simpler Lie algebras associated with the quantum Hamiltonian. First, by putting $\tau = t$ and $w = H = H(q, -i\hbar\, \partial/\partial q)$ in (2.101), it is evident that the dynamical evolution of all observables, as prescribed by (2.81), can be viewed as a canonical transformation parametrized by time and generated by the Hamiltonian. Canonical transformations generated by the Lie algebra $\mathcal{A}_H{}^*$ are especially noteworthy, since (2.115) shows that the Hamiltonian is invariant for them,

$$\mathbf{H}_\alpha \equiv \mathbf{H} \qquad \text{for} \quad \mathcal{A}_H{}^* \equiv \{\alpha \cdot w : [H, w_i] = 0\}, \qquad (2.121)$$

and (2.117) becomes

$$\dot{\mathbf{f}}_\alpha = [\mathbf{f}_\alpha, \mathbf{H}]. \qquad (2.122)$$

The latter is a generic dynamical equation identical to (2.81), and thus the dynamics of observables is unchanged by canonical transformations generated by $\mathcal{A}_H{}^*$. For this reason, the Lie algebra $\mathcal{A}_H{}^*$ is referred to as the *symmetry algebra* for the dynamical system, and the associated Lie group $\mathcal{G}_H{}^*$, with its linear representation on the space of wave functions $\mathcal{G}_H{}^* = \{(\exp \alpha \cdot \mathbf{G})\}$ provided by (2.113), is called the *symmetry group*. In the case of stationary states (2.61), the canonical transformation equation (2.118) gives

$$u_\alpha(q) = (\exp \alpha \cdot \mathbf{G})u(q), \qquad (2.123)$$

and group representation theory can be evoked to classify degenerate energy multiplets of stationary states.[20] Finally, we note the especially simple kind of dynamics that arises if $\mathbf{H}$ is an element of the Lie algebra $\mathscr{A}_s$ discussed above, that is, if H has the form of the right-hand side of (1.27). Because of the isomorphism of $\mathscr{A}_s$ to $\mathscr{A}_s$, the quantum dynamics associated with an $\mathbf{H}$ in $\mathscr{A}_s$ is identical to the classical dynamics associated with the corresponding H in $\mathscr{A}_s$ for corresponding observables in $\mathscr{A}_s$ and $\mathscr{A}_s$. In particular, the components of $\mathbf{q}(t)$ are inhomogeneous linear expressions in the components of $\mathbf{q}(0)$ and $\mathbf{p}(0)$ with the same time-dependent coefficients as in the classical theory, and likewise for the components of $\mathbf{p}(t)$. Dynamical containment in $\mathscr{A}_s$ is featured by all elements in the Lie algebra if $\mathbf{H}$ is in $\mathscr{A}_s$,

$$f(\mathbf{q}(t), \mathbf{p}(t)) = \tfrac{1}{2}(\mathbf{q}(0) \cdot \alpha(t) \cdot \mathbf{q}(0) + \mathbf{q}(0) \cdot \beta(t) \cdot \mathbf{p}(0)$$

$$+ \mathbf{p}(0) \cdot \beta'(t) \cdot \mathbf{q}(0) + \mathbf{p}(0) \cdot \gamma(t) \cdot \mathbf{p}(0)) + \xi(t) \cdot \mathbf{q}(0)$$

$$+ \eta(t) \cdot \mathbf{p}(0) + \omega(t), \tag{2.124}$$

where the components of $\beta'(t)$ are $\beta'_{ij}(t) \equiv \beta_{ji}(t)$ and with $\alpha(t)$, $\beta(t)$, $\gamma(t)$, $\xi(t)$, $\eta(t)$, and $\omega(t)$ the same as in (1.64), because of (2.81), the Lie algebra addition and Dirac bracket closure properties, and the isomorphism of $\mathscr{A}_s$ to $\mathscr{A}_s$.

---

[20] A lucid introduction to the application of symmetry group representation theory has been given by K. W. McVoy, *Rev. Modern Phys.* **37**, 84, 1965. For a more advanced account of the theory and recent applications, see F. J. Dyson, "Symmetry Groups." Benjamin, New York, 1966.

Chapter 3 *Classical Field Theory*

1. Neologized Lagrange Formulation

Taken as a primitive physical notion, the *state* of a continuous dynamical system with an infinite number of degrees of freedom is realized at any instant of time by fixing the components of $\phi = \phi(x) = (\phi_1(x), \ldots, \phi_n(x))$, the *field amplitude* real[1] *n-tuple* and $\dot\phi = \dot\phi(x) = (\dot\phi_1(x), \ldots, \dot\phi_n(x))$, the *field velocity* real[1] *n-tuple*. Here, $x = (x_1, x_2, x_3)$ denotes the Cartesian coordinates of a point in real physical space, and the components of ϕ and $\dot\phi$ are real continuous functions of x. No special transformation character is assumed to be featured by the field amplitude and field velocity *n*-tuples.

The system *dynamics*, again a primitive physical notion, is described by a function of time $\phi = \phi(x; t)$. This gives the evolution of the field amplitude *n*-tuple; the field velocity *n*-tuple is then required to take the associated form $\dot\phi = \partial\phi/\partial t$. Finally, physically admissible fields $\phi = \phi(x; t)$ are prescribed by a *dynamical law* in the form of *Euler–Lagrange field equations*

$$\delta L/\delta\phi(x) - (\partial/\partial t)\,(\delta L/\delta\dot\phi(x)) = 0, \tag{3.1}$$

---

[1] There is no loss of generality in taking the field amplitude and field velocity *n*-tuples to be real-valued, for by assigning distinct enumerator indices to the real and imaginary parts of complex-valued fields, the latter can be reexpressed and treated in terms of real quantities.

where the *Lagrangian* $L = L[\phi, \dot{\phi}]$ is a real scalar (1-tuple) functional of $\phi = \phi(x)$ and $\dot{\phi} = \dot{\phi}(x)$; in (3.1), n-tuple functional differential operators (see Appendix A) appear as

$$\delta/\delta\phi(x) = (\delta/\delta\phi_1(x), \ldots, \delta/\delta\phi_n(x)),$$
$$\delta/\delta\dot{\phi}(x) = (\delta/\delta\dot{\phi}_1(x), \ldots, \delta/\delta\dot{\phi}_n(x)). \tag{3.2}$$

The field theory is *local* if it is possible to express the Lagrangian as[2]

$$L = \int \mathcal{L}\, dx, \qquad dx \equiv dx_1\, dx_2\, dx_3, \tag{3.3}$$

where the *Lagrangian density* $\mathcal{L} = \mathcal{L}(\phi, \nabla\phi, \dot{\phi})$ is a real scalar function of ϕ and its first derivatives, the gradient of ϕ with respect to x being denoted by $\nabla\phi$. For a local field theory, the Euler–Lagrange field equations (3.1) become

$$(\partial\mathcal{L}/\partial\phi) - \sum_{i=1}^{3} \nabla_i(\partial\mathcal{L}/\partial\nabla_i\phi) - (\partial/\partial t)(\partial\mathcal{L}/\partial\dot{\phi}) = 0. \tag{3.4}$$

Consider, for example, the Lagrangian density

$$\mathcal{L} = \tfrac{1}{2}\, \dot{\phi} \cdot \dot{\phi} - \tfrac{1}{2} \sum_{i=1}^{3} \nabla_i\phi \cdot \nabla_i\phi - u, \tag{3.5}$$

where a dot between n-tuples denotes contraction and the *self-interaction energy density* $u = u(\phi)$ is a real 1-tuple function of ϕ; putting (3.5) into (3.4) produces

$$\ddot{\phi} - \nabla^2\phi + \partial u/\partial\phi = 0. \tag{3.6}$$

If $u = u(\phi)$ is just quadratic in ϕ, then (3.6) is a system of linear partial differential equations, and the field theory is referred to as *linear* (as well as local); the general solution to Eqs. (3.6) is readily obtainable for a linear field theory by Fourier analysis or other methods based on the principle of superposition of elementary solutions. The field theory is said to be *nonlinear* if u is not merely a quadratic form in ϕ and the system of partial differential equations (3.6) is nonlinear in ϕ. If u differs so significantly from a quadratic form in ϕ that the equations (3.6)

---

[2] The range for integration over x in (3.3) and subsequent equations is understood to be all of physical space, represented mathematically either as the unbounded 3-dimensional Euclidean space R_3 or as the limit $\mathcal{V} \to R_3$ of a large region $\mathcal{V} \subset R_3$ with periodic boundary conditions on the field amplitude and field velocity n-tuples.

are not amenable to approximate (perturbative) methods of solution (which employ linear analytical techniques based on the principle of superposition), the field theory is said to be *essentially nonlinear*. In any case, the Euler–Lagrange field equations (3.6) are *semilinear* because the terms in second partial derivatives of ϕ (the highest order derivative structure) occur linearly. Dynamical coupling between the component space-time functions of the field amplitude n-tuple is effected by the $\partial u/\partial \phi$ term in (3.6) for a self-interaction energy density such that $\partial^2 u/\partial \phi_i \, \partial \phi_j \neq 0$ for certain index values $i \neq j$.

In general, Eqs. (3.1) are invariant with respect to transformations of the form

$$L \rightarrow \left(L + \int \delta\chi/\delta\phi(x) \cdot \dot{\phi}(x)\, dx\right), \qquad (3.7)$$

where $\chi = \chi[\phi(x)]$ is an arbitrary functional of $\phi(x)$; the Euler–Lagrange field equations (3.4) for a local theory are invariant with respect to transformations of the form

$$\mathscr{L} \rightarrow (\mathscr{L} + \delta\chi/\delta\phi(x) \cdot \dot{\phi}(x)), \qquad (3.8)$$

where now the functional χ must be restricted to a form that does not involve spatial gradients of $\phi(x)$.[3] A given set of Euler–Lagrange field equations (3.1) or (3.4) requires a specific Lagrangian, unique to within a transformation (3.7) with Lagrangians related by such a transformation being "dynamically equivalent." In analogy to the quantity in classical mechanics, the *canonical energy* integral to (3.1) is

$$E \equiv \int \dot{\phi}(x) \cdot \frac{\delta L}{\partial \dot{\phi}(x)}\, dx - L. \qquad (3.9)$$

The latter quantity is constant in time, $\dot{E} = 0$ for $\phi = \phi(x; t)$ satisfying (3.1), and takes the same form for all Lagrangians which are related by a transformation (3.7). That no *inequivalent Lagrangians* [unrelated by transformations of the form (3.7)] produce the same prescribed set of Euler–Lagrange field equations can be demonstrated by exhibiting the

---

[3] The field equations (3.4) would not be affected if a pure spatial divergence were also to appear on the right-hand side of (3.8); we neglect such a term here because it would not modify the Lagrangian (3.3).

functional differential equations analogous to Eqs. (1.6) and (1.7) in classical mechanics and noting the overdetermined character of this system of equations for L; solutions [exemplified by (1.9)] for inequivalent Lagrangians that appear in classical mechanics (for exceptional dynamical systems with a single degree of freedom) have no analog for continuous dynamical systems with an infinite number of degrees of freedom.

The *action functional* is defined as

$$S = S[\phi(x; t)] \equiv \int_{t'}^{t''} L[\phi, \dot{\phi}]\, dt \qquad (3.10)$$

on the domain of real n-tuple functions $\phi(x; t)$ of continuity class C^1 with respect to t that reduce to the prescribed boundary values $\phi(x; t') = \phi'(x)$ and $\phi(x; t'') = \phi''(x)$, two fixed n-tuple functions of x. The difference of any two $\phi(x; t)$ in the domain of S is a real n-tuple function in the space $\mathscr{F} \equiv \{\sigma(x; t) : \sigma(x; t) \in C^1$ with respect to t for $t' \leqslant t \leqslant t''$, $\sigma(x; t')$ $= \sigma(x; t'') = 0\}$. For any $\sigma(x; t) \in \mathscr{F}$, we have

$$\left(\frac{d}{d\varepsilon} S[\phi(x; t) + \varepsilon\sigma(x; t)]\right)_{\varepsilon=0} \equiv \left(\frac{d}{d\varepsilon} \int_{t'}^{t''} L[\phi + \varepsilon\sigma, \dot{\phi} + \varepsilon\dot{\sigma}]\, dt\right)_{\varepsilon=0}$$

$$= \int_{t'}^{t''} \int \left(\sigma \cdot \frac{\delta L}{\delta \phi(x)} + \dot{\sigma} \cdot \frac{\delta L}{\delta \dot{\phi}(x)}\right) dx\, dt$$

$$= \int_{t'}^{t''} \int \sigma(x; t) \cdot \left(\frac{\delta L}{\delta \phi(x)} - \frac{\partial}{\partial t} \frac{\delta L}{\delta \dot{\phi}(x)}\right) dx\, dt,$$

$$(3.11)$$

while the functional derivatives of S with respect to components of ϕ follow from the implicit definition given in Appendix A as

$$\left((d/d\varepsilon)S[\phi(x; t) + \varepsilon\sigma(x; t)]\right)_{\varepsilon=0} \equiv \int_{t'}^{t''} \int \sigma(x; t) \cdot [\delta S/\delta\phi(x; t)]\, dx\, dt.$$

$$(3.12)$$

Hence, by comparing (3.11) and (3.12), we find that the functional derivatives of S with respect to the components of $\phi(x; t)$ are given explicitly as

$$\delta S/\delta\phi(x; t) = \delta L/\delta\phi(x) - (\partial/\partial t)(\delta L/\delta\dot{\phi}(x)), \qquad (3.13)$$

where the right-hand side of (3.13) is understood to be evaluated at $\phi(x) = \phi(x; t)$ and $\dot\phi(x) = \partial\phi(x; t)/\partial t$. Therefore, the Euler–Lagrange field equations (3.1) are equivalent to the *action principle*: S *has an extremum at a physical* $\phi(x; t)$,

$$\delta S/\delta\phi(x; t) = 0, \qquad (3.14)$$

with the left-hand side evaluated at a physical $\phi(x; t)$. The action principle (3.14) constitutes the most compact and fundamental statement of the dynamical law.

The existence and nature of singularity-free global solutions to nonlinear local classical field equations (such as (3.6) with $u = u(\phi)$ not merely a quadratic form in ϕ) has been the subject of numerous recent investigations[4] and is a topic of central importance for the future development of fundamental classical dynamical theory. Elementary analysis employing dilatation (scale) transformations of the field amplitude and its space-time arguments produces a necessary condition for the existence of spatially localized singularity-free periodic solutions,[5] as shown in Appendix I for the generic local field theory based on the Lagrangian density (3.5). With regard to the paramount issue of actually determining the manifold of singularity-free global solutions to nonlinear classical field equations, modern analytical,[6] topological,[7] and group-theoretic[8] methods of integration offer promise not afforded by more specialized classical techniques.[9] The dilatation transformation group-theoretic method for obtaining rigorous solutions to essentially nonlinear classical field equations is illustrated in Appendix J, where we apply the general theory to classical field equations of the form (3.6).

---

[4] See, for example, K. Jörgens, *Math. Z.* **77**, 295 (1961); I. Segal, *Proc. Symp. Appl. Math.* **17**, 210 (1965); C. S. Morawetz, *Proc. Roy. Soc. (London), Ser. A* **306**, 291 (1968).

[5] G. Rosen, *J. Math. Phys.* **7**, 2071 (1966).

[6] R. Courant and D. Hilbert, "Methods of Mathematical Physics II." Wiley (Interscience), New York, 1962; G. Rosen, *Phys. Rev.* **183**, 1186 (1969).

[7] M. A. Krasnosel'skii, "Topological Methods in the Theory of Nonlinear Integral Equations." Macmillan, New York, 1964; "Positive Solutions of Operator Equations." P. Noordhoff, Groningen, Netherlands, 1964.

[8] G. Birkhoff, "Hydrodynamics," Chaps. 4 and 5. Princeton Univ. Press, Princeton, New Jersey, 1960; A. J. A. Morgan, *Quart. J. Math., Oxford Ser.* **2**, 250 (1952).

[9] A. R. Forsyth, "Theory of Differential Equations," Vols. V and VI. Dover, New York, 1959.

2. Neologized Hamilton Formulation

Although the action principle (3.14) is the most fundamental statement of the dynamical law, an important subsidiary formulation follows directly from (3.1). First, the *field momentum density* real n-tuple

$$\pi = \pi(x) \equiv \delta L/\delta\dot{\phi}(x) \qquad (3.15)$$

is introduced, and the *Hamiltonian*

$$H = H[\phi, \pi] \equiv \int \dot{\phi}(x) \cdot \pi(x)\, dx - L = \left(\int \dot{\phi}(x) \cdot [\delta/\delta\dot{\phi}(x)]\, dx - 1 \right) L \qquad (3.16)$$

is defined, a linear functional Legendre transform of the Lagrangian expressed in terms of ϕ and π. The expression for the total variation

$$\delta H = \int (\dot{\phi}(x) \cdot \delta\pi(x) - [\delta L/\delta\phi(x)] \cdot \delta\phi(x))\, dx \qquad (3.17)$$

follows by evoking the definition (3.15). In combination with the Euler–Lagrange field equations (3.1), (3.17) implies that the *Hamiltonian canonical field equations*

$$\dot{\phi}(x) = \delta H/\delta\pi(x), \qquad \dot{\pi}(x) = -\delta H/\delta\phi(x) \qquad (3.18)$$

are satisfied by $\phi(x) = \phi(x; t)$ and $\pi(x) = \pi(x; t)$. Thus, for example, the Lagrangian density (3.5) produces the Hamiltonian

$$H = \int \left(\tfrac{1}{2}\pi \cdot \pi + \tfrac{1}{2} \sum_{i=1}^{3} \nabla_i \phi \cdot \nabla_i \phi + u \right) dx \qquad (3.19)$$

and associated canonical field equations

$$\dot{\phi}(x) = \pi(x), \qquad \dot{\pi}(x) = \nabla^2 \phi(x) - \partial u/\partial\phi(x).$$

In the Hamilton formulation with the $2n$ first-order field equations (3.18), a state of the system is realized at any instant of time by fixing the *canonical field variables* ϕ and π, the field amplitude and momentum density n-tuples. Evaluated at a solution to the field equations (3.18), the Hamiltonian (3.16) equals the canonical energy (3.9) and is usually[10] interpreted as the physical energy of the system.

---

[10] An example of the exceptional circumstance for which the evaluated Hamiltonian does not equal the physical energy appears in G. Wentzel, "Quantum Theory of Fields," p. 69. Wiley (Interscience), New York, 1949.

As in classical mechanics for systems with a finite number of degrees of freedom, the Lagrange formulation is not completely equivalent to the Hamilton formulation for fields. We have the Lagrangian given as a linear functional Legendre transform of the Hamiltonian,

$$L = \int \pi(x) \cdot \dot{\phi}(x) \, dx - H = \left(\int \pi(x) \cdot [\delta/\delta\pi(x)] \, dx - 1 \right) H,$$

$$(3.20)$$

as a consequence of (3.16) and the first canonical field equation (3.18). The "inverse" Hamiltonian-to-Lagrangian mapping (3.20) relinquishes dynamical effects associated with a term of first-order homogeneity in $\pi(x)$ in the Hamiltonian. For example, if the Hamiltonian (3.19) is modified by an additive term of first-order homogeneity in $\pi(x)$ to give

$$H = \int \left(\tfrac{1}{2}\pi \cdot \pi + \tfrac{1}{2} \sum_{i=1}^{3} \nabla_i \phi \cdot \nabla_i \phi + u + g\phi \cdot \pi \right) dx \qquad (3.21)$$

with g a constant parameter, the associated canonical field equations

$$\dot{\phi}(x) = \pi(x) + g\phi(x), \qquad \dot{\pi}(x) = \nabla^2\phi(x) - \partial u/\partial\phi(x) - g\pi(x) \quad (3.22)$$

combine to yield the classical field equations

$$\ddot{\phi}(x) - \nabla^2\phi(x) + \partial u/\partial\phi(x) - g^2\phi(x) = 0, \qquad (3.23)$$

while the Lagrangian density (3.5), associated with (3.21) by (3.20) and (3.3), produces the Euler–Lagrange field equations (3.6) without the term proportional to g^2 that appears in (3.23); instead, the field equations (3.23) follow from the (essentially unique) Lagrangian

$$L = \int \left(\tfrac{1}{2}\dot{\phi} \cdot \dot{\phi} - \tfrac{1}{2} \sum_{i=1}^{3} \nabla_i \phi \cdot \nabla_i \phi - u + \tfrac{1}{2}g^2\phi \cdot \phi \right) dx, \qquad (3.24)$$

which is canonically related [either directly by (3.16) or reciprocally by (3.20)] to the Hamiltonian

$$H = \int \left(\tfrac{1}{2}\pi \cdot \pi + \tfrac{1}{2} \sum_{i=1}^{3} \nabla_i \phi \cdot \nabla_i \phi + u - \tfrac{1}{2}g^2\phi \cdot \phi \right) dx, \qquad (3.25)$$

but no Lagrangian that produces the classical field equations (3.23) is related to the Hamiltonian (3.21). Conversely, although the Legendre transform (3.16) maps dynamically equivalent Lagrangians (related by a transformation (3.7)) into the same Hamiltonian, the canonical field equations in the Hamilton formulation are inequivalent to the Euler–Lagrange field equations if the Lagrangian features a term of first-order homogeneity in $\dot{\phi}$ which is not simply a linear functional in $\dot{\phi}$. Finally, if the Hamiltonian (3.16) vanishes identically for a certain Lagrangian, we have

$$\frac{dH}{dt} = \int \dot{\phi}(x) \cdot \left(-\frac{\delta L}{\delta \phi(x)} + \frac{\partial}{\partial t} \frac{\partial L}{\partial \dot{\phi}(x)} \right) dx \equiv 0;$$

hence, the associated Euler–Lagrange field equations (3.1) are not deterministic because a free constant of integration must appear in a solution $\phi(x; t)$ for prescribed initial data $[\phi(x; 0), \dot{\phi}(x; 0)]$.

Let us consider the Hamilton formulation correspondent of the action principle (3.14). In place of the action functional (3.10) associated with the Lagrange formulation, we must employ the *canonical field variable action functional*

$$\bar{S} = \bar{S}[\phi(x; t), \pi(x; t)] \equiv \int_{t'}^{t''} \left(\int \dot{\phi}(x) \cdot \pi(x) \, dx - H[\phi, \pi] \right) dt,$$

$$(3.26)$$

obtained by using (3.16) to eliminate the Lagrangian from (3.10) and defined on the domain of real n-tuple functions $\phi(x; t)$ of continuity class C^1 with respect to t that reduce to the prescribed boundary values $\phi(x; t') = \phi'(x)$ and $\phi(x; t'') = \phi''(x)$, fixed n-tuple functions of x, and $\pi(x; t)$ of continuity class C^0 with respect to t (unconstrained at $t = t'$ and $t = t''$). The $2n$ first-order Hamilton canonical field equations (3.18) are equivalent to the *canonical field variable action principle*:

$$\delta \bar{S}/\delta \phi(x; t) = 0, \qquad \delta \bar{S}/\delta \pi(x; t) = 0, \qquad (3.27)$$

where the functional derivatives are taken according to the general definition in Appendix A with $\mathcal{F} = \{\sigma(x; t) : \sigma(x, t) \in C^1$ with respect to t for $t' \leqslant t \leqslant t''$, $\sigma(x; t') = \sigma(x; t'') = 0\}$ in the case of the $\phi(x; t)$ differentiation and with $\mathcal{F} = \{\sigma(x; t) : \sigma(x; t) \in C^0$ with respect to t for $t' \leqslant t \leqslant t''\}$ in the case of the $\pi(x; t)$ differentiation. With these definitions of functional differentiation, the canonical field variable action principle (3.27) states that $\bar{S}$ has an extremum for a physical pair of n-tuple functions $\phi(x; t)$, $\pi(x; t)$. Because $\phi(x; t)$ and $\pi(x; t)$ play asymmetric roles in (3.26) and (3.27), the canonical field variable action principle does not provide a logical starting point for a discussion of canonical transformations.

3. Neologized Poisson Formulation

A useful feature of the Hamilton canonical field equations (3.18) is that they allow the time rate of change of any *observable* $f = f[\phi, \pi]$, defined here as an infinitely differentiable real 1-tuple functional in the canonical field variables, to be expressed directly in terms of ϕ and π as

$$df/dt \equiv \dot{f} = [f, H] \tag{3.28}$$

with the field-theoretic Poisson bracket of two observables introduced as

$$[f, g] \equiv \int \left(\frac{\delta f}{\delta \phi(x)} \cdot \frac{\delta g}{\delta \pi(x)} - \frac{\delta g}{\delta \phi(x)} \cdot \frac{\delta f}{\delta \pi(x)} \right) dx. \tag{3.29}$$

For example, consider the observable

$$f = \iint (\tfrac{1}{2}\phi(x) \cdot \alpha(x, y) \cdot \phi(y) + \phi(x) \cdot \beta(x, y) \cdot \pi(y)$$

$$+ \tfrac{1}{2}\pi(x) \cdot \gamma(x, y) \cdot \pi(y)) \, dx \, dy$$

$$+ \int (\xi(z) \cdot \phi(z) + \eta(z) \cdot \pi(z)) \, dz + \omega, \tag{3.30}$$

where $\alpha(x, y)$, $\beta(x, y)$, and $\gamma(x, y)$ denote real n^2-dyad distributions (generalized functions) in x, y with $\alpha(x, y)$ and $\gamma(x, y)$ symmetrical under a simultaneous interchange of their (continuous) arguments and (suppressed) indices, $\xi(z)$ and $\eta(z)$ denote real n-tuple distributions in z, and ω denotes a real 1-tuple constant; the Poisson bracket of the observable (3.30) and the Hamiltonian observable[11] (3.19) is the observable

$$[f, H] = \iint \left(\phi(x) \cdot \alpha(x, y) \cdot \pi(y) + \pi(x) \cdot \beta(x, y) \cdot \pi(y) \right.$$

$$+ [\phi(x) \cdot \beta(x, y) + \pi(x) \cdot \gamma(x, y)]$$

$$\left. \cdot \left[\nabla^2 \phi(y) - \frac{\partial u(\phi(y))}{\partial \phi(y)} \right] \right) dx \, dy$$

$$+ \int \left(\xi(z) \cdot \pi(z) + \eta(z) \cdot \left[\nabla^2 \phi(z) - \frac{\partial u(\phi(z))}{\partial \phi(z)} \right] \right) dz.$$

---

[11] For the Hamiltonian (3.19) to be an infinitely differentiable functional in the canonical field variables, u must be an infinitely differentiable function of ϕ.

In particular, the observable constants of the motion are quantities which have a zero Poisson bracket with the Hamiltonian.

The set of all observables is closed under ordinary addition and Poisson bracket combination, in the sense that both the sum and the Poisson bracket of any two observables are observables. Furthermore, the Poisson bracket combination law (3.29), associating a new observable functional of ϕ and π with an ordered pair of observable functionals of ϕ and π, has all the properties required of a combination law for a Lie algebra, properties (1.28)–(1.30). Hence, the set of all observable functionals subject to ordinary addition and Poisson bracket combination (3.29) constitutes a Lie algebra, and any subset of observable functionals that is closed with respect to ordinary addition and Poisson bracket combination consitutes a Lie subalgebra of the Lie algebra of all observable functionals. The subset of observables having the generic form (3.30) constitutes an infinite-dimensional Lie subalgebra because an infinite number of real constants are required to specify $\alpha(x, y)$, $\beta(x, y)$, and so forth. In subsequent discussion, we refer to the set of all observable functionals having the generic form (3.30) as the infinite-dimensional Lie algebra $\mathfrak{A}_s$, the field-theoretic analog of the $(2n^2 + 3n + 1)$-dimensional Lie algebra of classical mechanics denoted by $\mathscr{A}_s$. The main structure of $\mathfrak{A}_s$ is evident in the subset with $\xi(z) = \eta(z) \equiv 0$, $\omega = 0$; this subset of observable functionals, which we denote by $\mathfrak{A}_s'$, is closed with respect to ordinary addition and Poisson bracket combination, and hence constitutes an infinite-dimensional Lie subalgebra of $\mathfrak{A}_s$; the subset with $\alpha(x, y) = \beta(x, y) = \gamma(x, y) \equiv 0$ also constitutes an infinite-dimensional Lie subalgebra of $\mathfrak{A}_s$, which we denote by $\mathfrak{A}_s''$. Because the Poisson bracket of an observable in $\mathfrak{A}_s'$ and an observable in $\mathfrak{A}_s''$ is an observable in $\mathfrak{A}_s''$, $\mathfrak{A}_s$ is the semidirect sum of $\mathfrak{A}_s'$ and $\mathfrak{A}_s''$,

$$\mathfrak{A}_s = \mathfrak{A}_s' \oplus \mathfrak{A}_s''.$$

To the Lie algebra $\mathscr{A}$ of a subset of observable functionals is associated a conjugate Lie algebra $\mathscr{A}^*$ consisting of all observable functionals which have a zero Poisson bracket with all elements of $\mathscr{A}$. As in classical mechanics, the observable constants of the motion constitute the Lie algebra $\mathscr{A}_H^*$, conjugate to the 1-dimensional Lie algebra consisting of elements proportional to the Hamiltonian, $\mathscr{A}_H \equiv \{(\text{real constant parameter}) \times H\}$. However, in contrast to classical mechanics where $(2n - 1)$

functionally independent constants of the motion are generally associated with the Hamiltonian as a complete set of solutions to the first-order
linear homogeneous partial differential equations $[f_i, H] = 0$, in classical field theory there is no established theorem which gives the number
of functionally independent constants of the motion, independent of
the detailed form of the Hamiltonian. For a generic linear field theory,
the first-order linear homogeneous functional differential equations
$[f_i, H] = 0$ admit an infinite number of functionally independent
observable functions f_i as constants of the motion, exemplified by[12]

$$f_i = \int (\pi \cdot (-\nabla^2)^i \pi + \phi \cdot (-\nabla^2)^{i+1} \phi) \, dx \qquad (i = 0, \pm 1, \pm 2, \ldots)$$
$$(3.31)$$

in the case of the simple linear field theory with

$$H = \int \left(\tfrac{1}{2} \pi \cdot \pi + \tfrac{1}{2} \sum_{i=1}^{3} \nabla_i \phi \cdot \nabla_i \phi \right) dx. \qquad (3.32)$$

Recent work[13] supports the conjecture that a generic nonlinear field
theory also admits an infinite number of functionally independent
observables as constants of the motion.[14] Associated with the infinite-
dimensional Lie algebra $\mathscr{A}_H{}^*$ [which contains infinitely differentiable
functions of (functionally independent) observable constants of the
motion], we have the Lie algebra $\hat{\mathscr{A}}_H{}^*$ consisting of a maximal set, closed
under Poisson bracket combination, of linearly independent constants
of the motion (exclusive of H and the trivial generic observables that are
constants of the motion for any H, as exemplified by $\int \phi \cdot \nabla_i \pi \, dx$ for
$i = 1, 2, 3$). For a generic linear field theory, $\hat{\mathscr{A}}_H{}^*$ is an infinite-dimensional Lie algebra, as exemplified for the simple linear theory with the
Hamiltonian (3.32) by the set (3.31) (the element $i = 0$ being excluded),
for which $[f_i, f_j] = 0$. It is likely that $\hat{\mathscr{A}}_H{}^*$ is infinite-dimensional for a
generic nonlinear field theory. Major importance would be attached to
a general proof of the infinite dimensionality of $\hat{\mathscr{A}}_H{}^*$ and to a general
algorithm for constructing its elements.[14]

---

[12] Negative powers of the negative Laplacian are defined by inserting δ-functions
and using $(-\nabla^2)^{-1} \delta(x) = (2\pi)^{-3} \int (k \cdot k)^{-1} e^{ik \cdot x} \, dk = (4\pi)^{-1} (x_1^2 + x_2^2 + x_3^2)^{-1/2}$.

[13] P. D. Lax, *Comm. Pure Appl. Math.* **21**, 467 (1968).

[14] A general algorithm for the construction of such observable functionals would
serve to elucidate the nature of singularity-free global solutions to nonlinear classical
field equations; what is needed is a general integration theory for first-order linear
homogeneous functional differential equations of the form $[f_i, H] = 0$ with H
prescribed.

A transformation of all observables

$$f = f[\phi, \pi] \overset{w}{\to} f_\tau = f_\tau[\phi, \pi]$$

$$g = g[\phi, \pi] \overset{w}{\to} g_\tau = g_\tau[\phi, \pi] \qquad (3.33)$$

$$\cdots$$

such that

$$f_0 = f, \qquad g_0 = g, \qquad \cdots \qquad (3.34)$$

is said to be a *canonical transformation* generated by the observable $w = w[\phi, \pi]$ if the transformed observables satisfy the first-order total differential equations

$$df_\tau/d\tau = [f_\tau, w]$$

$$dg_\tau/d\tau = [g_\tau, w] \qquad (3.35)$$

$$\cdots$$

A transformed observable is given explicitly in terms of the original observable by the Maclaurin series derived from (3.34) and (3.35),

$$f_\tau = f + \tau[f, w] + \frac{\tau^2}{2!} [[f, w], w] + \cdots$$

$$\equiv \left(\exp \tau \int \left(\frac{\delta w}{\delta \pi(x)} \cdot \frac{\delta}{\delta \phi(x)} - \frac{\delta w}{\delta \phi(x)} \cdot \frac{\delta}{\delta \pi(x)} \right) dx \right) f. \qquad (3.36)$$

As readily verified by direct computation, we have

$$h = [f, g] \Rightarrow h_\tau = [f_\tau, g_\tau] \qquad (3.37)$$

for $\tau \geq 0$, so canonical transformations in classical field theory preserve the Lie algebraic structure for a subset of observable functionals that is closed with respect to addition and Poisson bracket combination: Lie algebras are mapped into isomorphic Lie algebras by canonical transformations. By putting $h = f$ and $g = H$ in (3.37), it follows that the time rate of change of any canonically transformed observable is prescribed by a dynamical equation of the form (3.28),

$$df_\tau/dt \equiv \dot{f}_\tau = [f_\tau, H_\tau]. \qquad (3.38)$$

By specializing (3.37) to the cases for which $h = c$, a quantity independent of ϕ and π, we find that

$$[f, g] = c \Rightarrow [f_\tau, g_\tau] = c. \qquad (3.39)$$

In particular, for the components of the field amplitude and momentum density n-tuples, we have the basic Poisson bracket relations

$$[\phi_i(x), \phi_j(y)] = [\pi_i(x), \pi_j(y)] = 0,$$
$$[\phi_i(x), \pi_j(y)] = \delta_{ij}\,\delta(x - y), \tag{3.40}$$

so that (3.39) yields

$$[\phi_{\tau i}(x), \phi_{\tau j}(y)] = [\pi_{\tau i}(x), \pi_{\tau j}(y)] = 0, \tag{3.41}$$

$$[\phi_{\tau i}(x), \pi_{\tau j}(y)] = \delta_{ij}\delta(x - y). \tag{3.42}$$

If two observables f and g are regarded as functionals of the canonically transformed components of the field amplitude and momentum density n-tuples, their Poisson bracket computed with respect to ϕ_τ and π_τ equals their Poisson bracket computed with respect to ϕ and π,

$$\int \left(\frac{\delta f}{\delta \phi_\tau(x)} \cdot \frac{\delta g}{\delta \pi_\tau(x)} - \frac{\delta g}{\delta \phi_\tau(x)} \cdot \frac{\delta f}{\delta \pi_\tau(x)} \right) dx$$

$$= \int \left(\frac{\delta f}{\delta \phi(x)} \cdot \frac{\delta g}{\delta \pi(x)} - \frac{\delta g}{\delta \phi(x)} \cdot \frac{\delta f}{\delta \pi(x)} \right) dx, \tag{3.43}$$

the proof being analogous to the proof for (1.45) in classical mechanics. Hence, the Poisson bracket in an equation can be computed with respect to ϕ_τ, π_τ for any value of $\tau \geqslant 0$. By doing this in (3.38) and specializing to the cases for which f_τ equals each of the $2n$ components of ϕ_τ and π_τ, we see that canonical transformations also preserve the form of the Hamilton canonical field equations,

$$\dot{\phi}_\tau(x) = \delta H_\tau / \delta \pi_\tau(x), \qquad \dot{\pi}_\tau(x) = -\delta H_\tau / \delta \phi_\tau(x). \tag{3.44}$$

Under canonical transformations generated by an m-dimensional Lie algebra $\mathscr{A}$ composed of all real linear combinations of m linearly independent observable functionals $w_1, \ldots, w_m$, a generic observable $f = f[\phi, \pi]$ is transformed to

$$f_\alpha = (\exp \alpha \cdot G)f, \tag{3.45}$$

in which the generators for canonical transformations

$$G_i \equiv \int \left(\frac{\delta w_i}{\delta \pi(x)} \cdot \frac{\delta}{\delta \phi(x)} - \frac{\delta w_i}{\delta \phi(x)} \cdot \frac{\delta}{\delta \pi(x)} \right) dx \qquad (3.46)$$

are m first-order functional differential operators that satisfy commutation relations of the form

$$G_i G_j - G_j G_i = \sum_{k=1}^{m} c_{ijk} G_k . \qquad (3.47)$$

All properties derived for the transformed observables (3.36) are featured by the transformed observables (3.45). As an illustration of canonical transformations generated by a finite-dimensional Lie algebra, consider the 3-dimensional subalgebra of $\mathfrak{A}_s$,

$$\mathscr{A} = \left\{ \alpha_1 w_1 + \alpha_2 w_2 + \alpha_3 w_3 : w_1 \equiv \tfrac{1}{4} \int (\pi(x) \cdot \pi(x) - \phi(x) \cdot \phi(x)) \, dx, \right.$$

$$w_2 \equiv -\tfrac{1}{4} \int (\phi(x) \cdot \phi(x) + \pi(x) \cdot \pi(x)) \, dx,$$

$$\left. w_3 \equiv -\tfrac{1}{2} \int \phi(x) \cdot \pi(x) \, dx \right\}. \qquad (3.48)$$

For this $\mathscr{A}$ the generators (3.46) are

$$G_1 = \tfrac{1}{2} \int (\phi(x) \cdot \delta/\delta\pi(x) + \pi(x) \cdot \delta/\delta\phi(x)) \, dx$$

$$G_2 = \tfrac{1}{2} \int (\phi(x) \cdot \delta/\delta\pi(x) - \pi(x) \cdot \delta/\delta\phi(x)) \, dx \qquad (3.49)$$

$$G_3 = \tfrac{1}{2} \int (\pi(x) \cdot \delta/\delta\pi(x) - \phi(x) \cdot \delta/\delta\phi(x)) \, dx$$

and satisfy (3.47) with $c_{123} = c_{231} = -c_{312} = 1$; thus, (3.45) with (3.49) gives the effect of $SL(2, R)$ canonical transformations on the space of observable functionals.

It is evident that the dynamical evolution of all observables, as prescribed by (3.28), can be viewed as a canonical transformation parametrized by $\tau = t$ with $f_t = f_t[\phi(x; 0), \pi(x; 0)] \equiv f[\phi(x; t), \pi(x; t)]$ and generated by the Hamiltonian with $w = H[\phi(x; 0), \pi(x; 0)]$. This special canonical transformation takes $\phi(x; 0) \overset{H}{\rightarrow} \phi_t(x; 0) \equiv \phi(x; t)$ and $\pi(x; 0) \overset{H}{\rightarrow} \pi_t(x; 0) \equiv \pi(x; t)$, with the Poisson bracket in (3.35) computed

most conveniently with respect to the "original" canonical field variables ($\phi(x; 0)$, $\pi(x; 0)$). The dynamical evolution of all observables is
expressed in the canonical transformation form (3.36) as

$$f = f[\phi(x; t), \pi(x; t)] = \left(\exp t \int \left(\frac{\delta H}{\delta \pi(x)} \cdot \frac{\delta}{\delta \phi(x)} - \frac{\delta H}{\delta \phi(x)} \cdot \frac{\delta}{\delta \pi(x)}\right) dx\right)$$

$$\times f[\phi(x), \pi(x)]\Big|_{\substack{\phi(x) = \phi(x; 0) \\ \pi(x) = \pi(x; 0)}} \tag{3.50}$$

Since the Lie algebra $\mathscr{A}_H{}^*$ is infinite-dimensional for a generic linear
field theory and conjectured to be infinite-dimensional for a generic nonlinear field theory, the notion of canonical transformations generated by
an m-dimensional Lie algebra must be extended to the case of an
infinite-dimensional Lie algebra in order to apply to an $\mathscr{A}_H{}^*$ in classical
field theory. We have $G_i H = 0$ for the generators associated with $\mathscr{A}_H{}^*$,
and thus the Hamiltonian is invariant for canonical transformations
generated by $\mathscr{A}_H{}^*$ for α such that the sum in (3.45), $\alpha \cdot G \equiv \sum_{i=1}^{\infty} \alpha_i G_i$,
converges. The Lie algebra $\mathscr{A}_H{}^*$ is referred to as the *symmetry algebra*
for the classical field theory because the dynamics of observables is
unchanged by canonical transformations generated by $\mathscr{A}_H{}^*$ with
$H_\alpha = H$, and the associated Lie group $\mathscr{G}_H{}^*$, with its linear representation
on the space of observable functionals $\mathscr{G}_H{}^* = \{(\exp \alpha \cdot G)\}$, is called the
symmetry group.

An especially simple kind of dynamics arises if H is an element of the
Lie algebra $\mathfrak{A}_s$, that is, if H has the form of the right-hand side of (3.30),
as it does generally for any linear field theory with the Euler–Lagrange
field equations (3.1) linear and homogeneous in $\phi = \phi(x, t)$. [For
example, the Hamiltonian (3.32) is the element of $\mathfrak{A}_s$ with $\alpha(x, y) =$
$-\nabla^2 \delta(x - y)\mathbf{1}$, $\beta(x, y) = 0$, $\gamma(x, y) = \delta(x - y)\mathbf{1}$, $\xi(z) \equiv 0$, $\eta(z) \equiv 0$,
$\omega = 0$.] Because the components of $\phi(x)$ and $\pi(x)$ are elements of $\mathfrak{A}_s{}''$
(with $\alpha(x, y) = \beta(x, y) = \gamma(x, y) \equiv 0$, $\omega = 0$, and one component of
either $\xi(z)$ or $\eta(z)$ equal to $\delta(x - z)$, all other components of $\xi(z)$ and
$\eta(z) \equiv 0$) and their Poisson brackets with elements of $\mathfrak{A}_s$ are elements of
$\mathfrak{A}_s{}''$, it follows from (3.45) that elements of $\mathfrak{A}_s$ generate canonical transformations for which the components of $\phi_\alpha(x)$ are inhomogeneous linear
expressions in the components of $\phi(x)$ and $\pi(x)$, and likewise for the
components of $\pi_\alpha(x)$. In particular, the dynamical evolution generated
by an H in $\mathfrak{A}_s$ is characterized by the components of $\phi(x; t)$ being inhomogeneous linear expressions in the components of $\phi(x; 0)$ and

$\pi(x; 0)$, and likewise for the components of $\pi(x; t)$. Moreover, if H is an element of $\mathfrak{A}_s$, all observables in $\mathfrak{A}_s$ at $t = 0$ remain in $\mathfrak{A}_s$ for $t > 0$, and any observable that is not initially in $\mathfrak{A}_s$ does not enter the algebra. Dynamical containment in $\mathfrak{A}_s$ is featured by all elements in the Lie algebra if H is in $\mathfrak{A}_s$,

$$f[\phi(x; t), \pi(x; t)]$$

$$\equiv f_t[\phi(x; 0), \pi(x; 0)]$$

$$= \iint (\tfrac{1}{2}\phi(x; 0) \cdot \alpha(x, y; t) \cdot \phi(y; 0) + \phi(x; 0) \cdot \beta(x, y; t) \cdot \pi(y; 0)$$

$$+ \tfrac{1}{2}\pi(x; 0) \cdot \gamma(x, y; t) \cdot \pi(y; 0))\, dx\, dy + \int (\xi(z; t) \cdot \phi(z; 0)$$

$$+ \eta(z; t) \cdot \pi(z; 0))\, dz + \omega(t), \tag{3.51}$$

because of the Lie algebra addition and Poisson bracket closure properties.

Chapter 4 *Quantum Field Theory*

1. Neologized Feynman Formulation

Again taken as a primitive physical notion, the *state* of a continuous dynamical system with an infinite number of degrees of freedom is realized at any instant of time by prescribing the complex *wave functional* $\Psi = \Psi[\phi]$ that depends on the field amplitude real n-tuple $\phi = \phi(x) = (\phi_1(x), \ldots, \phi_n(x))$. The associated positive-definite quantity $|\Psi[\phi]|^2$ is interpreted as the relative *probability density* for a physical measurement fixing the state at the field amplitude n-tuple $\phi = \phi(x)$. This role assigned to a complex wave functional is the basic postulate in quantum field theory.

The system *dynamics* is described by assigning the wave functional a time-dependent form $\Psi = \Psi[\phi; t]$ (C^1 with respect to t). A physically admissible evolution in time takes the specific *linear* form of a *dynamical principle of superposition*

$$\Psi[\phi''; t''] = \int K[\phi'', \phi'; t'' - t']\Psi[\phi'; t']\mathscr{D}(\phi') \tag{4.1}$$

for all $t'' \geqslant t'$ and initial wave functionals $\Psi = \Psi[\phi'; t']$; in (4.1), the integration is over all $\phi' = \phi'(x)$, the ϕ'-space infinitesimal volume element is represented symbolically as

$$\mathscr{D}(\phi') = \mathbf{N} \prod_{\text{all } x} \prod_{i=1}^{n} d\phi_i'(x) \tag{4.2}$$

68

with the components of $x = (x_1, x_2, x_3)$ taking on all real values in the infinite product, and the normalization constant **N** is independent of ϕ'. Viewed physically in terms of a sufficiently accurate finite-dimensional approximation to ϕ'-space, the infinitesimal volume element (4.2) is a displacement-invariant Haar measure, like $D(\hat{q})$ in the propagation kernel representation (2.13), but with the integration variable ϕ' in $\mathscr{D}(\phi')$ parametrized by the three real coordinates in x.[1] To guarantee the general consistency of (4.1) for all $\Psi[\phi'; t']$, the *propagation kernel* $K[\phi'', \phi'; t'' - t']$ must satisfy a *semigroup composition law*

$$K[\phi'', \phi'; t'' - t'] = \int K[\phi'', \phi; t'' - t] K[\phi, \phi'; t - t'] \mathscr{D}(\phi) \quad (4.3)$$

for $t' \leqslant t \leqslant t''$, as well as the initial value condition

$$\lim_{t \to 0} K[\phi'', \phi'; t] = \delta[\phi'' - \phi'] \equiv \mathbf{N}^{-1} \prod_{\text{all } x} \prod_{i=1}^{n} \delta(\phi_i(x)), \quad (4.4)$$

where the δ-*functional* on the right-hand side of (4.4) is such that

$$\delta[\phi'' - \phi'] = 0 \qquad \text{for} \quad \phi'' \neq \phi',$$

$$\int \delta[\phi] \mathscr{D}(\phi) = 1. \quad (4.5)$$

Repeated application of (4.3) shows that the propagation kernel for finite values of $(t'' - t')$ can be expressed as an iteration of the propagation kernel for infinitesimal values of $(t'' - t')$,

$K[\phi'', \phi'; t'' - t']$

$$= \int \left(\prod_{M=1}^{N} K[\phi(x; t' + M \Delta t), \phi(x; t' + (M-1) \Delta t); \Delta t] \right)$$

$$\times \prod_{M=1}^{N-1} \mathscr{D}(\phi(x; t' + M \Delta t)), \quad (4.6)$$

---

[1] Precise criteria for existence and methods of construction of displacement-invariant measure for integration over fields have been given by K. O. Friedrichs and H. Shapiro, "Seminar on Integration of Functionals." Courant Institute of Mathematical Sciences, New York Univ., 1957; K. O. Friedrichs and H. N. Shapiro, *Proc. Nat. Acad. Sci.* **43**, 336 (1957); N. N. Sudakow, *Dokl. Akad. Nauk SSSR* **127**, 524 (1959); B. S. Mitjagin, *Uspekhi Math. Nauk* **26**, 191 (1961). The general formulations of displacement-invariant measure on infinite-dimensional topological function spaces have been reviewed by N. Dunford and J. Schwartz, "Linear Operators, I," p. 402 ff. Wiley (Interscience), New York, 1958; E. J. McShane, *Bull. Amer. Math. Soc.* **69**, 597 (1963); J. M. Gelfand and N. J. Wilenkin, "Verallgemeinerte Funktionen, IV," *Hochschubücher Math. Bd.* **50**, (1964).

where the notation is analogous to that employed in writing (2.6) with the time variable in $\phi(x; t)$ serving as a virtually continuous parameter for the intermediate integrations between t' and t'' as $\Delta t \equiv [(t'' - t')/N] \to 0$, the fixed n-tuple field amplitude arguments in the propagation kernel being prescribed by

$$\phi(x; t') \equiv \phi'(x) \qquad \text{and} \qquad \phi(x; t'') \equiv \phi''(x). \tag{4.7}$$

In analogy to Feynman's prescription (2.7), the infinitesimal propagation kernel in (4.6) is postulated to have the form

$$K[\phi(x; t + \Delta t), \phi(x; t); \Delta t] = A^{-1} \exp(iL[\phi, \dot\phi] \, \Delta t/\hbar) \tag{4.8}$$

in which

$$\phi = \tfrac{1}{2}[\phi(x; t + \Delta t) + \phi(x; t)], \qquad \dot\phi = (\Delta t)^{-1}[\phi(x; t + \Delta t) - \phi(x; t)], \tag{4.9}$$

and the normalization constant $A = A(\Delta t)$. The Feynman *functional integral representation*[2]

$$K[\phi'', \phi'; t'' - t'] = \int_{\mathscr{C}} (\exp iS/\hbar) \mathfrak{D}(\phi) \tag{4.10}$$

follows in the limit $N \to \infty$, $\Delta t \to 0$ of (4.6) with (4.8), where the action functional $S = S[\phi(x; t)]$ defined by (3.10) appears in the form of a Riemann sum, and $\mathfrak{D}(\phi)$ in (4.10) denotes an infinitesimal volume element for the integration over the class of real n-tuple functions $\mathscr{C} = \{\phi = \phi(x; t) \text{ for } t' \leqslant t \leqslant t'': \phi(x; t') = \phi'(x), \phi(x; t'') = \phi''(x)\}$. In symbolic notation, we have

$$\mathfrak{D}(\phi) = \mathscr{N} \prod_{t' < t < t''} \prod_{\text{all } x} \prod_{i=1}^{n} d\phi_i(x; t) \tag{4.11}$$

with the generic normalization factor

$$\mathscr{N} = \mathscr{N}(t'' - t') \equiv \lim_{\Delta t \to 0} ([A(\Delta t)]^{-(t'' - t')/\Delta t} \mathbf{N}^{(t'' - t' - \Delta t)/\Delta t}) \tag{4.12}$$

independent of $\phi(x; t)$. By introducing

$$\hat\phi = \hat\phi(x; t) \equiv \phi(x; t) - [((t - t')\phi''(x) + (t'' - t)\phi'(x))/(t'' - t')], \tag{4.13}$$

---

[2] R. P. Feynman, *Phys. Rev.* **80**, 440 (1950); **84**, 108 (1951).

as have $\mathfrak{D}(\phi) = \mathfrak{D}(\hat{\phi})$, and thus (4.10) becomes

$$K[\phi'', \phi'; t'' - t'] = \int_{\mathscr{G}} (\exp iS/\hbar)\mathfrak{D}(\hat{\phi}) \qquad (4.14)$$

with the integration over the space of real n-tuple functions

$$\mathscr{G} = \{\hat{\phi} = \hat{\phi}(x; t) \quad \text{for} \quad t' \leqslant t \leqslant t'' : \hat{\phi}(x; t') = 0 = \hat{\phi}(x; t'')\}$$

Displacement-invariant for any fixed σ in $\mathscr{G}$,

$$\mathfrak{D}(\hat{\phi} + \sigma) = \mathfrak{D}(\hat{\phi}), \qquad (4.15)$$

the infinitesimal volume element in (4.14) can be viewed physically as a left-invariant and right-invariant Haar measure for the integration over a large but finite-dimensional Abelian topological subgroup of $\mathscr{G}$, like $D(\hat{q})$ in (2.13).[1] Hence, the functional integral representation of the propagation kernel (4.14) is defined uniquely to within a normalization factor independent of ϕ'' and ϕ' because of the general Haar measure existence and uniqueness theorems. Formally similar to the normalization factor (2.11) in the functional integral for the quantum mechanics propagation kernel (2.13), the normalization factor (4.12) in (4.11) is fixed to within a complex constant (associated with a renormalization transformation of the form (2.15)) by the physically required existence of the propagation kernel (4.14) for all $t'' \geqslant t'$, by the semigroup composition law (4.3), and by the initial value condition (4.4).

Equations (4.1) and (4.14) constitute the Feynman "sum-over-histories" formulation of the dynamical law for a continuous system with an infinite number of degrees of freedom. The theory would suffice in a very elegant way for closed nondissipative quantum fields associated with a classical Lagrangian if we could evaluate the functional integral (4.14) for all physically important actions $S - S[\phi(x; t)]$. However, the functional integral representation of the propagation kernel (4.14) is not amenable to rigorous evaluation except for an action that is a *general quadratic* expression in $\phi(x; t)$ (possibly including a linear functional term), for which the Lagrangian functional in (3.10) can be expressed generically as

$$L[\phi, \dot{\phi}] = \iint (\tfrac{1}{2}\dot{\phi}(x) \cdot \rho(x, y) \cdot \dot{\phi}(y) + \tfrac{1}{2}\dot{\phi}(x) \cdot \theta(x, y) \cdot \phi(y)$$

$$+ \tfrac{1}{2}\phi(x) \cdot \zeta(x, y) \cdot \phi(y)) \, dx \, dy - \int \xi(z) \cdot \phi(z) \, dz,$$

$$(4.16)$$

modulo an additive real constant and a transformation of the form (3.7) with χ a general quadratic functional of $\phi(x)$; in (4.16), $\rho(x, y)$, $\theta(x, y)$, and $\zeta(x, y)$ denote real n^2-dyad distributions (generalized functions) in x, y with $\rho(x, y)$ and $\zeta(x, y)$ symmetrical, $\theta(x, y)$ antisymmetrical, under a simultaneous interchange of their (continuous) arguments and (suppressed) indices, and $\xi(z)$ denotes a real n-tuple distribution in z; it should be noted that a Lagrangian of the form (4.16) generally produces nonlocal, inhomogeneous, linear Euler–Lagrange field equations (3.1). If $S = S[\phi(x; t)]$ is a general quadratic, then we have

$$S[\phi(x; t)] = S[\phi_c(x; t)] + S[\tilde{\phi}(x; t)] \tag{4.17}$$

where

$$\tilde{\phi}(x; t) \equiv \phi(x; t) - \phi_c(x; t) \qquad \text{with} \quad \phi_c(x; t)$$

the classical "history" satisfying the action principle (3.14),

$$\delta S/\delta\phi(x; t)\Big|_{\phi=\phi_c} = 0, \tag{4.18}$$

or equivalently, with the Lagrangian given by (4.16),

$$\int (\rho(x, y) \cdot \ddot{\phi}_c(y; t) + \theta(x, y) \cdot \dot{\phi}_c(y; t)$$

$$- \zeta(x, y) \cdot \phi_c(y; t)) \, dy + \xi(x) = 0, \tag{4.19}$$

subject to the boundary conditions $\phi_c(x; t') = \phi'(x)$ and $\phi_c(x; t'') = \phi''(x)$. With $\phi(x; t)$ an element of $\mathscr{C}$, it follows that the variable n-tuple function $\tilde{\phi}(x; t)$ is an element of $\mathscr{G}$, that is, $\tilde{\phi}(x; t') = \tilde{\phi}(x; t'') = 0$. Hence, by setting the fixed σ in (4.15) equal to

$$\sigma(x; t) = [((t - t')\phi''(x) + (t'' - t)\phi'(x))/(t'' - t')] - \phi_c(x; t), \tag{4.20}$$

we find that the displacement-invariant measure $\mathfrak{D}(\tilde{\phi}) = \mathfrak{D}(\phi)$. Therefore, (4.14) produces the result

$$K[\phi'', \phi'; t'' - t'] = \mathscr{N}(\exp iS[\phi_c(x; t)]/\hbar) \tag{4.21}$$

with the factor

$$\int_{\mathscr{G}} (\exp iS[\tilde{\phi}(x; t)]/\hbar)\mathfrak{D}(\tilde{\phi})$$

independent of ϕ', ϕ'' and absorbed into (4.12) to give the normalization factor $\mathscr{N}$ in (4.21). As a prime example, the Lagrangian

$$L = \int \left(\tfrac{1}{2}\, \dot{\phi} \cdot \dot{\phi} - \tfrac{1}{2}\, (\nabla\phi) \cdot (\nabla\phi) - \tfrac{1}{2} \sum_{i=1}^{n} m_i^2 \phi_i^2 \right) dx, \qquad (4.22)$$

which specializes (4.16) to a local linear theory of uncoupled field amplitude components, leads to a propagation kernel of the form (4.21) with

$$S[\phi_c(x; t)]$$

$$= \int \sum_{i=1}^{n} (\tfrac{1}{2}\phi_i''(x)(-\nabla^2 + m_i^2)^{1/2}(\tan[(-\nabla^2 + m_i^2)^{1/2}(t'' - t')])^{-1}\phi_i''(x)$$

$$+ \tfrac{1}{2}\phi_i'(x)(-\nabla^2 + m_i^2)^{1/2}(\tan[(-\nabla^2 + m_i^2)^{1/2}(t'' - t')])^{-1}\phi_i'(x)$$

$$- \phi_i''(x)(-\nabla^2 + m_i^2)^{1/2}(\sin[(-\nabla^2 + m_i^2)^{1/2}(t'' - t')])^{-1}\phi_i'(x)) \, dx.$$

$$(4.23)$$

The normalization factor $\mathscr{N} = \mathscr{N}(t'' - t')$ associated with (4.22) and (4.23) cannot be expressed as a finite function of $(t'' - t')$, as is generally the case for such a normalization factor in the Feynman formulation of quantum field theory. However, since the state functional at any instant of time is physically significant modulo a complex constant independent of ϕ, this feature of the formulation does not impair practical calculations. The standard procedure is to require normalization of the wave functional at the time of interest, $\int |\Psi[\phi; t]|^2 \mathscr{D}(\phi) = 1$, concomitant with the definition of expectation values (defined by (4.54) below) and other quantities amenable to experimental measurement at time t.

As in quantum mechanics with the Feynman functional integral representation for the propagation kernel (2.13), we can extract useful dynamical information from (4.14) for quantum fields by asymptotic techniques and other suitable analytical procedures if $S = S[\phi(x; t)]$ is not a general quadratic expression that can be evaluated explicitly. For example, if the action functional takes the form $S = S_0 + S_{\text{int}}$, where $S_0 = S_0[\phi(x; t)]$ is simple [that is, a general quadratic functional in $\phi(x; t)$] and $S_{\text{int}} = S_{\text{int}}[\phi(x; t)]$ is relatively small compared to S_0 for typical $\phi(x; t)$, then the propagation kernel can be evaluated approximately by working out the leading terms in the expansion of (4.14),

$$K[\phi'', \phi'; t'' - t'] = \sum_{k=0}^{\infty} (k!)^{-1}(i/\hbar)^k \int_{\mathscr{G}} (S_{\text{int}})^k (\exp iS_0/\hbar) \mathfrak{D}(\hat{\phi}). \quad (4.24)$$

The latter perturbation expansion, directly related to the conventional *iteration solution* for the wave functional in the Schrödinger formulation [as shown by formula (4.52)], provides an immediate conceptual basis for a large fraction of the calculations performed in quantum field theory. In particular, Feynman's important work on quantum electro-dynamics[3] stemmed from considerations based on formula (4.24).[4] Applications of the expansion (4.24) lead one to study *Feynman operators* in field theory, functional integrals of the form

$$\int_{\mathscr{G}} A(\exp iS/\hbar)\mathfrak{D}(\hat{\phi}) \equiv (\phi''; t'' |A[\phi(x; t)]| \phi'; t')_S, \qquad (4.25)$$

where $S = S[\phi(x; t)]$ and $A = A[\phi(x; t)]$ are generic real-valued functionals of $\phi(x; t)$. The "functional integration by parts" lemma derived in Appendix D applies to such functional integrals in the theory with displacement-invariant Haar measure, and it allows one to express *exact* equations involving Feynman operators without having to do any explicit functional integration. For example, we obtain the *operator Euler–Lagrange field equations*

$$\int_{\mathscr{G}} \frac{\delta S}{\delta\phi(x; t)} (\exp iS/\hbar)\mathfrak{D}(\hat{\phi})$$

$$\equiv \left(\phi''; t'' \left| \frac{\delta S}{\delta\phi(x; t)} \right| \phi'; t'\right)_S$$

$$= 0 \qquad \text{for all} \quad x \quad \text{and} \quad t' < t < t'', \qquad (4.26)$$

and the Feynman operator equations

$$\int_{\mathscr{G}} \left[\frac{\delta S}{\delta\phi_i(x; t)} \phi_j(y; s) - i\hbar\, \delta_{ij}\, \delta(x - y)\, \delta(t - s) \right] (\exp iS/\hbar)\mathfrak{D}(\hat{\phi})$$

$$\equiv \left(\phi''; t'' \left\| \left[\frac{\delta S}{\delta\phi_i(x; t)} \phi_j(y; s) - i\hbar\, \delta_{ij}\, \delta(x - y)\, \delta(t - s) \right] \right\| \phi'; t'\right)_S$$

$$= 0 \qquad \text{for all} \quad x, y \quad \text{and} \quad t' < t, \quad s < t''.$$

$$(4.27)$$

[3] R. P. Feynman, "Quantum Electrodynamics." Benjamin, New York, 1961, and papers reprinted therein.

[4] Interesting recent work based on the Feynman perturbation expansion (4.24) has been published by E. S. Fradkin, *Nucl. Phys.* **49**, 624 (1963); *Acta Phys. Hung.* **19**, 175 (1965); C. S. Lam, *Nuovo Cimento* **50A**, 504 (1967).

As in quantum mechanics, exact equations such as (4.26) and (4.27) facilitate the determination of Feynman operators and the computation of terms in (4.24) without explicit functional integration. Thus, for example, if the action is the time integral of (4.22), Eqs. (4.27) yield

$$\left(-\frac{\partial^2}{\partial t^2} + \nabla^2 - m_i^2\right)(\phi''; t'' \,|\phi_i(x; t)\phi_j(y; s)|\, \phi'; t')_S$$

$$= i\hbar\, \delta_{ij}\, \delta(x - y)\, \delta(t - s)(\phi''; t'' \,|1|\, \phi'; t')_S, \tag{4.28}$$

in which the propagation kernel (4.21) with (4.23) appears as the Feynman operator $(\phi''; t''\, |1|\, \phi'; t')_S$ on the right-hand side; the *causal Green's function* of perturbation theory is obtained by integrating the vacuum expectation value of (4.28) subject to appropriate boundary conditions,[5] obviating calculation of the causal Green's function by explicit functional integration.[6] It is also noteworthy that the connection between Feynman operators and the more conventional differential operators of quantum field theory [expressed by Eq. (4.53)] is readily established as the analog of the relationship derived in Appendix F for the operators in quantum mechanics; hence, Feynman operator equations, such as (4.26), (4.27), and (4.28), imply corresponding Heisenberg operator equations for the field variables.

The canonical variable version of the "sum-over-histories" functional integral representation for the propagation kernel follows in quantum field theory as

$$K[\phi'', \phi'; t'' - t'] = \int_{\mathscr{C}\times\mathscr{F}} (\exp i\bar{S}/\hbar)\mathfrak{D}(\phi, \pi), \tag{4.29}$$

where the canonical field variable action functional

$$\bar{S} = \bar{S}[\phi(x; t), \pi(x; t)]$$

defined by (3.26) appears in the form of a Riemann sum, and $\mathfrak{D}(\phi, \pi)$ in (4.29) denotes an infinitesimal volume element for integration over the class of real n-tuple functions

$$\mathscr{C} = \{\phi = \phi(x; t) \quad \text{for} \quad t' \leqslant t \leqslant t'' : \phi(x; t') = \phi'(x),\, \phi(x; t'') = \phi''(x)\}$$

and the space of real n-tuple functions

$$\mathscr{F} = \{\pi = \pi(x; t) \quad \text{for} \quad t' \leqslant t \leqslant t''\},$$

[5] C. S. Lam, *Nucl. Phys.* **87**, 549 (1967).
[6] P. T. Matthews and A. Salam, *Nuovo Cimento* **2**, 120 (1955).

the field momentum density *n*-tuple being unconstrained at $t = t'$ and $t = t''$. In symbolic notation, we have

$$\mathfrak{D}(\phi, \pi) = \left(\prod_{t' < t < t''} \prod_{\text{all } x} \prod_{i=1}^{n} d\phi_i(x; t) \right) \left(\prod_{t' \leqslant t \leqslant t''} \mathbf{N} \prod_{\text{all } x} \prod_{i=1}^{n} d\pi_i(x; t) \right)$$

(4.30)

with a generic normalization constant **N** appearing in the infinite product of momentum density infinitesimal volume elements; thus, the measure in (4.29) is *displacement-invariant* for any fixed $\sigma \in \mathcal{G} = \{\hat{\phi} = \hat{\phi}(x; t) \text{ for } t' \leqslant t \leqslant t'' : \hat{\phi}(x; t') = 0 = \hat{\phi}(x; t'')\}$ and $\omega \in \mathcal{F}$,

$$\mathfrak{D}(\phi + \sigma, \pi + \omega) = \mathfrak{D}(\phi, \pi).$$

(4.31)

It follows from (4.31) that the Feynman functional integral representation (4.29) yields the representation (4.14) for field theories described by a Hamiltonian of the form

$$H = \int (\tfrac{1}{2}\pi \cdot \pi + a \cdot \pi) \, dx + V$$

(4.32)

where $a = a(\phi, \nabla\phi, \ldots)$ is a real *n*-tuple function of ϕ and spatial derivatives of ϕ, and $V = V[\phi]$ is a real 1-tuple functional of ϕ (that is, $\delta V/\delta\pi(x) \equiv 0$). More generally, for a Hamiltonian which is not of the form (4.32), the dynamical laws (4.14) and (4.29) are inequivalent. Thus, for the Hamiltonian

$$H = \int (\pi_1 \nabla^2 \phi_2 - \pi_2 \nabla^2 \phi_1) \, dx$$

(4.33)

straightforward evaluation of (4.29) yields the propagation kernel

$$K[\phi'', \phi'; t'' - t'] = \delta[\Gamma\phi'' - \phi'],$$

$$(\Gamma\phi'')_1 \equiv (\cos[(t'' - t') \nabla^2])\phi_1'' - (\sin[(t'' - t') \nabla^2])\phi_2'' \quad (4.34)$$

$$(\Gamma\phi'')_2 \equiv (\sin[(t'' - t') \nabla^2])\phi_1'' + (\cos[(t'' - t') \nabla^2])\phi_2'',$$

while (4.14) is ill-defined with the Lagrangian obtained by putting (4.33) into (3.20) identically zero.[7] Both Feynman functional integral

---

[7] If, however, we disregard the $\dot{\pi}_1, \dot{\pi}_2$ canonical field equations derived from (4.33), the $\dot{\phi}_1, \dot{\phi}_2$ equations follow from the (essentially unique but canonically unrelated) Lagrangian

$$L = \tfrac{1}{2} \int (\dot{\phi}_1 \phi_2 - \dot{\phi}_1 \phi_2 - (\nabla\phi_1)^2 - (\nabla\phi_2)^2) \, dx$$

for which (4.14) yields a propagation kernel of the form (4.21), well-defined but distinctly unequal to (4.34).

representations exist and are unequal for certain canonically related Lagrangians and Hamiltonians. It is necessary to appeal to the observable physics in order to determine whether (4.14) or (4.29) is the appropriate dynamical law for such quantum field theories.

A formal quasiclassical approximation for the propagation kernel (4.14) is derived in the same fashion as (2.39) for the propagation kernel (2.13) in quantum mechanics. If the action evaluated at the classical "history" is large in absolute magnitude compared to $\hbar$, we obtain

$$K[\phi'', \phi'; t'' - t'] \cong \mathcal{N} \, [\det(\delta^2 S_c)]^{-1/2}(\exp iS[\phi_c(x; t)]/\hbar), \quad (4.35)$$

where $\mathcal{N} = \mathcal{N}(t'' - t')$ is a normalization factor (generally not expressible as a finite function in field theory), and $\det(\delta^2 S_c)$ denotes the infinite product of all eigenvalues λ_μ of the symmetric matrix kernel

$$\delta^2 S/\delta\phi(x; r) \, \delta\phi(y; s) \Big|_{\phi = \phi_c}; \quad \det(\delta^2 S_c) \equiv \prod_{\mu=1}^{\infty} \lambda_\mu$$

must exist modulo formal (not necessarily finite) normalization depending on $(t'' - t')$ for the quasiclassical approximation (4.35) to have applicability in quantum field theory.

2. Neologized Schrödinger Formulation

An alternative formulation of the dynamical law for a quantum field theory is obtained if (4.8) is used to derive a linear functional differential equation for the wave functional, rather than a functional integral representation for the propagation kernel. By differentiating (4.1) with respect to t'' and letting $t' \to t''$ in the integral, we find that

$$\partial\Psi[\phi''; t'']/\partial t'' = \int \Lambda[\phi'', \phi']\Psi[\phi'; t'']\mathcal{D}(\phi'), \quad (4.36)$$

in which

$$\Lambda[\phi'', \phi'] \equiv \lim_{\Delta t \to 0} \frac{\partial K[\phi'', \phi'; \Delta t]}{\partial(\Delta t)}$$

$$= \lim_{\Delta t \to 0} \left\{ \left[(i\hbar)^{-1}H - A^{-1}\frac{dA}{d(\Delta t)} \right] K[\phi'', \phi'; \Delta t] \right\}.$$

$$(4.37)$$

In (4.37), we have employed (4.8) and (4.9) with $\phi''(x)$ in place of $\phi(x; t + \Delta t)$ and $\phi'(x)$ in place $\phi(x; t)$, and thus the Hamiltonian (3.16) appears in terms of $\phi \equiv \frac{1}{2}(\phi'' + \phi')$ and $\dot{\phi} \equiv (\Delta t)^{-1}(\phi'' - \phi')$. To resolve the indeterminacy in (4.37) as $\Delta t \to 0$ (stemming from the Hamiltonian's dependency on $\dot{\phi}$), we note that

$$\frac{\delta}{\delta\phi''(x)} (\delta[\phi'' - \phi']) = \lim_{\Delta t \to 0} \frac{\delta K[\phi'', \phi'; \Delta t]}{\delta\phi''(x)}$$

$$= \lim_{\Delta t \to 0} \left\{ \frac{i(\Delta t)}{\hbar} \frac{\delta L[\phi, \dot{\phi}]}{\delta\phi''(x)} K[\phi'', \phi'; \Delta t] \right\}$$

$$= \frac{i}{\hbar} \left(\lim_{\Delta t \to 0} \pi(x) \right) \delta[\phi'' - \phi'], \tag{4.38}$$

where $\pi(x)$ is the field momentum density n-tuple (3.15), expressed in terms of $\phi = \frac{1}{2}(\phi'' + \phi')$ and $\dot{\phi} = (\Delta t)^{-1}(\phi'' - \phi')$. It follows from (4.38) that the field momentum density n-tuple multiplied by $\delta[\phi'' - \phi']$ is well defined as $\Delta t \to 0$ and given by

$$\left(\lim_{\Delta t \to 0} \pi(x) \right) \delta[\phi'' - \phi'] = -i\hbar[\delta/\delta\phi''(x)](\delta[\phi'' - \phi']), \tag{4.39}$$

but powers of $\pi_i(x) \equiv \delta L/\delta\dot{\phi}_i(x)$ in the Hamiltonian in (4.37) cannot be replaced simply by corresponding powers of $-i\hbar \, \delta/\delta\phi_i''(x)$, as evidenced by the terms $(i\hbar^{-1} \delta^2 L/\delta\phi_i(x) \, \delta\dot{\phi}_i(x) + i(\hbar \, \Delta t)^{-1} \delta^2 L/\delta\dot{\phi}_i(x)^2)$ in $K[\phi'', \phi'; \Delta t]^{-1} \delta^2 K[\phi'', \phi'; \Delta t]/\delta\phi_i''(x)^2$. It is these extra pure imaginary terms that counter the normalization term $A^{-1} \, dA/d(\Delta t)$ in (4.37) as $\Delta t \to 0$, so that subject to a suitable renormalization transformation (2.15), (4.37) yields

$$\Lambda[\phi'', \phi'] = (i\hbar)^{-1}H[\phi''(x), -i\hbar \, \delta/\delta\phi''(x)] \, \delta[\phi'' - \phi'] \tag{4.40}$$

for a certain symmetrical ordering of ϕ, π product terms in $H[\phi, \pi]$. Thus, the functional differential operator $H[\phi(x), -i\hbar \, \delta/\delta\phi(x)]$ is Hermitian on a space of sufficiently smooth complex-valued functionals of $\phi(x)$, functionals which are absolute-square-integrable over all $\phi(x)$ with the displacement-invariant measure (4.2) and a suitable rule for normalization. By putting (4.40) into (4.36), we obtain the Schrödinger equation for the wave functional

$$i\hbar \, (\partial\Psi[\phi; t]/\partial t) = H[\phi(x), -i\hbar \, \delta/\delta\phi(x)]\Psi[\phi; t]. \tag{4.41}$$

Note that the semigroup composition law (4.3) and the initial value condition (4.4) are sufficient to give the propagation kernel in terms of an unspecified *quantum Hamiltonian* $\mathbf{H}$ (namely, the generator associated with the one-parameter dynamical semigroup) as

$$K[\phi'', \phi'; t - t'] = (\exp - i(t - t')\mathbf{H}/\hbar)\, \delta[\phi'' - \phi'], \qquad (4.42)$$

the formal dynamical equation for the wave functional as

$$\Psi[\phi; t] = (\exp - i(t - t')\mathbf{H}/\hbar)\Psi[\phi; t'], \qquad (4.43)$$

and the *abstract Schrödinger equation* as

$$i\hbar(\partial\Psi[\phi; t]/\partial t) = \mathbf{H}\Psi[\phi; t], \qquad (4.44)$$

where $\mathbf{H}$ is a functional differential operator in $\phi(x)$ and $\delta/\delta\phi(x)$; the origin and structure of the quantum Hamiltonian

$$\mathbf{H} = H[\phi(x), -i\hbar\, \delta/\delta\phi(x)], \qquad (4.45)$$

the classical Hamiltonian suitably ordered to be Hermitian and with $\pi(x)$ replaced by $-i\hbar\, \delta/\delta\phi(x)$, is elucidated by the postulate for the infinitesimal propagation kernel (4.8).

The *iteration solution* for a wave functional satisfying Eq. (4.41) with $H = H_0 + H_{int}$, where $H_0 = H_0[\phi, \pi]$ is simple (that is, quadratic in the canonical field variables) and $H_{int} = H_{int}[\phi]$ is relatively small compared to H_0 for typical values of the canonical field variables, is derived in manifest analogy to the iteration solution for the wave function in quantum mechanics described in Appendix H. With

$$\mathbf{H}_0 \equiv H_0[\phi(x), -i\hbar\, \delta/\delta\phi(x)], \qquad (4.46)$$

the dynamical evolution of the *interaction wave functional*

$$\Psi_{int}[\phi; t] \equiv (\exp i\mathbf{H}_0 t/\hbar)\Psi[\phi; t] \qquad (4.47)$$

is given by

$$\Psi_{int}[\phi; t''] = \mathbf{U}(t'', t')\Psi_{int}[\phi; t'], \qquad (4.48)$$

$$\mathbf{U}(t'', t') = T\left(\exp - i \int_{t'}^{t''} H_{int}(\phi(x; t))\, dt/\hbar\right), \qquad (4.49)$$

in which the chronological ordering symbol T arranges noncommuting factors to the left with increasing values of t, and the operator field amplitude n-tuple is introduced as

$$\phi(x; t) \equiv (\exp i\mathbf{H}_0 t/\hbar)\phi(x)(\exp - i\mathbf{H}_0 t/\hbar). \qquad (4.50)$$

From (4.48) we obtain the iteration solution in its usual perturbation-theoretic form by expanding the *U-matrix* (4.49),

$$\mathbf{U}(t'', t') = \sum_{k=0}^{\infty} (k!)^{-1}(i/\hbar)^k T\left(- \int_{t'}^{t''} H_{int}(\boldsymbol{\phi}(x; t)) \, dt\right)^k. \quad (4.51)$$

The operator $\mathbf{U}(\infty, -\infty)$, which maps $\Psi_{int}[\phi; -\infty]$ into $\Psi_{int}[\phi; +\infty]$ according to (4.48), is referred to as the *S-matrix* in quantum field theory. Finally, by inverting the definition (4.47) and recalling the propagation kernel equation (4.1), we obtain the formula

$$K[\phi'', \phi'; t'' - t'] = \sum_{k=0}^{\infty} (k!)^{-1}(i/\hbar)^k(\exp -i\mathbf{H}_0''t''/\hbar)$$

$$\times T\left(- \int_{t'}^{t''} H_{int}(\boldsymbol{\phi}''(x; t)) \, dt\right)^k (\exp i\mathbf{H}_0''t'/\hbar)$$

$$\times \delta[\phi'' - \phi'], \quad (4.52)$$

where $\mathbf{H}_0''$ and $\boldsymbol{\phi}''(x; t)$ are defined by (4.46) and (4.50) with $\phi(x)$ replaced by $\phi''(x)$; formula (4.52) is equivalent to (4.24) because of the analog of (F.8) for Feynman operators in field theory,

$$\int_{\mathscr{G}} A(\exp iS_0/\hbar)\mathfrak{D}(\hat{\phi}) \equiv (\phi''; t'' |A[\phi(x; t)]| \phi'; t')_{S_0}$$

$$= (\exp -i\mathbf{H}_0''t''/\hbar)T(A[\boldsymbol{\phi}''(x; t)])(\exp i\mathbf{H}_0''t'/\hbar)$$

$$\times \delta[\phi'' - \phi']. \quad (4.53)$$

With the state of the system represented by a wave functional $\Psi[\phi; t]$, the *expectation value* of an observable $f[\phi(x), \pi(x)]$ is postulated as

$$\langle f \rangle \equiv \int \Psi[\phi; t]^* f[\phi(x), -i\hbar \, \delta/\delta\phi(x)]\Psi[\phi; t]\mathscr{D}(\phi) \quad (4.54)$$

for a suitable (Hermitian and experimentally appropriate) ordering of ϕ, π product terms, if such terms with noncommuting factors appear in $f[\phi(x), \pi(x)]$. As in previous functional integrals, $\mathscr{D}(\phi)$ in (4.54) denotes the displacement-invariant measure (4.2) for integration over all fields $\phi = \phi(x)$; the wave functional in (4.54) is assumed to be normalized to unity, $\int |\Psi[\phi; t]|^2 \mathscr{D}(\phi) = 1$. Again it is Eq. (4.39), derived from the Feynman formulation, that elucidates the origin and structure of the functional differential operator associated with an observable in (4.54).

Stationary states in quantum field theory are described by wave functionals of the form

$$\Psi[\phi; t] = (\exp -iEt/\hbar)U[\phi] \tag{4.55}$$

that satisfy (4.41), and hence $U[\phi]$ is an eigenfunctional solution to the Schrödinger stationary state equation

$$\mathbf{H}U[\phi] = H[\phi(x), -i\hbar\, \delta/\delta\phi(x)]U[\phi] = EU[\phi] \tag{4.56}$$

associated with the constant energy eigenvalue E. If the quantum Hamiltonian in (4.56) admits a complete orthonormal set of complex-valued eigenfunctionals $\{U_\mu[\phi]\}$ associated with the discrete set of energy eigenvalues $\{E_\mu\}$,[8] we have

$$\mathbf{H}U_\mu[\phi] = E_\mu\, U_\mu[\phi], \qquad \int U_\mu{}^*[\phi]U_\nu[\phi]\mathscr{D}(\phi) = \delta_{\mu\nu},$$
$$\sum_\mu U_\mu[\phi'']U_\mu{}^*[\phi'] = \delta[\phi'' - \phi']. \tag{4.57}$$

Then it follows that the propagation kernel in (4.1) has the formal representation

$$K[\phi'', \phi'; t'' - t'] = \sum_\mu (\exp -iE_\mu(t'' - t')/\hbar)U_\mu[\phi'']U_\mu{}^*[\phi'], \tag{4.58}$$

which satisfies the initial value condition (4.4) because of the completeness relation in (4.57) and is the formal solution to the propagation kernel equation implied by (4.41) and (4.1),

$$(i\hbar(\partial/\partial t'') - H[\phi''(x), -i\hbar\, \delta/\delta\phi''(x)])K[\phi'', \phi'; t'' - t'] = 0, \qquad (t'' > t'). \tag{4.59}$$

With only one eigenfunctional associated with the *vacuum state energy* $E_0 \equiv \min_\mu \{E_\mu\}$, (4.58) yields the formulas

$$E_0 = -\lim_{s\to\infty}(\hbar/s) \ln K[\phi'', \phi'; -is], \tag{4.60}$$

$$U_0[\phi'']U_0{}^*[\phi'] = \lim_{s\to\infty} \{(\exp E_0 s/\hbar)K[\phi'', \phi'; -is]\}. \tag{4.61}$$

---

[8] Ordinarily, the spectrum of the quantum Hamiltonian in field theory is continuous if x ranges over the unbounded three-dimensional Euclidean space R_3. A discrete spectrum of energy eigenvalues is usually induced if x ranges over a large region $\mathscr{V} \subset R_3$ with periodic boundary conditions on the field amplitude n-tuple; the summations over μ in Eqs. (4.57) and (4.58) are then interpreted in the symbolic sense (that is, representing integrations appropriately weighted with the number density of energy eigenvalues) in the limit $\mathscr{V} \to R_3$.

Similar relations for the E_μ and $U_\mu[\phi]$ of other stationary states are obtained by successively removing terms from (4.58) and rewriting (4.60) and (4.61) with the subtracted propagation kernels; thus, energy eigenvalues and eigenfunctionals can be extracted by simple limiting procedures from a closed-form expression for the propagation kernel, as provided by the Feynman functional integral representation (4.14). For example, from the propagation kernel (4.21) with (4.23), we obtain

$$E_0 = - \lim_{s \to \infty}((\hbar/s) \ln \mathscr{N}(-is)) \tag{4.62}$$

$$U_0[\phi'']U_0{}^*[\phi'] = \prod_{i=1}^{n} \left(\exp -(2\hbar)^{-1} \int (\phi_i''(x)(-\nabla^2 + m_i{}^2)^{1/2}\phi_i''(x) \right.$$

$$\left. + \phi_i'(x)(-\nabla^2 + m_i{}^2)^{1/2}\phi_i'(x) \right) dx \tag{4.63}$$

$$\Rightarrow U_0[\phi]$$

$$= \left(\begin{array}{c}\text{phase}\\\text{const}\end{array}\right) \prod_{i=1}^{n} \left(\exp -(2\hbar)^{-1} \int \phi_i(x)(-\nabla^2 + m_i{}^2)^{1/2}\phi_i(x) \, dx \right). \tag{4.64}$$

The Hamiltonian related canonically to (4.22) by (3.16) is

$$H[\phi, \pi] = \int \left(\tfrac{1}{2}\pi \cdot \pi + \tfrac{1}{2}(\nabla\phi) \cdot (\nabla\phi) + \tfrac{1}{2} \sum_{i=1}^{n} m_i{}^2\phi_i{}^2 \right) dx, \tag{4.65}$$

and direct computation shows that (4.64) is, indeed, the vacuum state solution to Eq. (4.56),

$$\int \left(-\frac{\hbar^2}{2} \frac{\delta}{\delta\phi(x)} \cdot \frac{\delta}{\delta\phi(x)} + \frac{1}{2}(\nabla\phi(x)) \cdot (\nabla\phi(x)) + \frac{1}{2} \sum_{i=1}^{n} m_i{}^2\phi_i(x)^2 \right) dx$$

$$\times U_0[\phi] = E_0 \, U_0[\phi], \tag{4.66}$$

with the vacuum state energy,

$$E_0 = \tfrac{1}{2}\hbar \left(\sum_{i=1}^{n} (-\nabla^2 + m_i{}^2)^{1/2} \, \delta(x) \right)_{x=0} \int dx$$

an infinite constant.[9] Not a measurable quantity in itself, the vacuum state energy provides the reference value for definition of the *observable energy* $(E_\mu - E_0)$ associated with the Hamiltonian eigenfunctional $U_\mu[\phi]$. In the case of Hamiltonians that feature a term quadratic in the momentum density *n*-tuple, as exemplified by (4.65), the stationary state eigenvalue equation $HU_\mu[\phi] = E_\mu U_\mu[\phi]$ is reduced to an eigenvalue equation for $(E_\mu - E_0)$ by putting $U_\mu[\phi] \equiv \Omega_\mu[\phi] U_0[\phi]$. As an example, for the Hamiltonian (4.65) we have

$$\int \sum_{i=1}^{n} \left(-\frac{\hbar^2}{2} \frac{\delta^2 \Omega_\mu}{\delta \phi_i(x)^2} + \hbar \phi_i(x)(-\nabla^2 + m_i^2)^{1/2} \frac{\delta \Omega_\mu}{\delta \phi_i(x)} \right) dx = (E_\mu - E_0)\Omega_\mu,$$

(4.67)

with the solutions for $\Omega_\mu = \Omega_\mu[\phi]$ readily obtainable as polynomial functionals in the field amplitude *n*-tuple,

$$\Omega_1 = \int \xi(x) \cdot \phi(x) \, dx$$

where $\xi(x)$ satisfies

$$\hbar(-\nabla^2 + m_i^2)^{1/2} \xi_i(x) = (E_1 - E_0)\xi_i(x), \qquad (4.68)$$

$$\Omega_2 = \frac{1}{2} \int \phi(x) \cdot \zeta(x, y) \cdot \phi(y) \, dx \, dy - \frac{\hbar^2}{2} (E_2 - E_0)^{-1} \int \sum_{i=1}^{n} \zeta_{ii}(x, x) \, dx,$$

where

$$\zeta_{ij}(x, y) = \zeta_{ji}(y, x),$$

$$[(-\nabla_x^2 + m_i^2)^{1/2} + (-\nabla_y^2 + m_j^2)^{1/2}] \zeta_{ij}(x, y) = (E_2 - E_0)\zeta_{ij}(x, y), \ \ldots .$$

(4.69)

Specialized to the $n = 3$ case with all m_i equal to zero, these stationary state solutions appear with slight modification in the quantum theory of electromagnetic radiation, outlined in Appendix K. Closely related stationary state solutions appear for certain quantum field theories of a general nonlinear character, as shown in Appendix L.

The Schrödinger formulation also supplies a dynamical description for quantum fields without a classical analog, systems that cannot be

---

[9] The fact that the quantity $\{\int U^*[\phi]HU[\phi]\mathscr{D}(\phi)/\int |U[\phi]|^2 \mathscr{D}(\phi)\}$ is stationary with respect to variations of $U[\phi]$ about solutions to Eq. (4.56) precludes existence of finite-energy eigenfunctionals for a large class of local field theories. For the method of proof, see G. Rosen, *J. Math. Phys.* **9**, 804 (1968).

treated by the Feynman formulation with a propagation kernel given by (4.14). Generally, the state of such a quantum field is described by an abstract Hilbert space vector $\mathbf{\Psi} = \mathbf{\Psi}(t)$ that satisfies the Schrödinger equation

$$i\hbar \, \partial\mathbf{\Psi}/\partial t = \mathbf{H}\mathbf{\Psi}, \qquad (4.70)$$

where the quantum Hamiltonian $\mathbf{H}$ must be prescribed without appeal to an equation of the form (4.45). For example, the field theory for neutrinos features the quantum Hamiltonian

$$\mathbf{H} = \int \psi^\dagger(q)\mathbf{D}\psi(q) \, dq \qquad (4.71)$$

with $\psi(x)$ a two-component field operator that satisfies the anti-commutation relations

$$\psi_i(x)\psi_j(y) + \psi_j(y)\psi_i(x) = 0$$
$$\psi_i(x)\psi_j{}^\dagger(y) + \psi_j{}^\dagger(y)\psi_i(x) = \delta_{ij} \, \delta(x - y) \qquad (4.72)$$
$$\psi_i{}^\dagger(x)\psi_j{}^\dagger(y) + \psi_j{}^\dagger(y)\psi_i{}^\dagger(x) = 0,$$

in which $\psi_i{}^\dagger(x)$ is the Hermitian adjoint of the operator $\psi_i(x)$ on the Hilbert space of $\mathbf{\Psi}$, $i, j = 1, 2$, and the matrix-differential operator $\mathbf{D}$ in (4.71) is the one-neutrino quantum Hamiltonian expressed by the right-hand side of (2.71). The field theory for electrons also features a quantum Hamiltonian of the form (4.71), but $\psi(x)$ a four-component field operator that satisfies anticommutation relations (4.72), $i, j = 1, 2, 3, 4$, the matrix-differential operator $\mathbf{D}$ in (4.71) being the one-electron quantum Hamiltonian expressed by the right-hand side of (2.72). Although it follows from (4.70) that the generic dynamical evolution of the state vector

$$\mathbf{\Psi}(t) = (\exp -i(t - t')\mathbf{H}/\hbar)\mathbf{\Psi}(t') \qquad (4.73)$$

is consistent with (4.43), a dynamical law of the form (4.1) and the concomitant notion of a propagation kernel do not apply to a quantum field without a classical analog. Hence, a Feynman formulation is not available for such a quantum field.[10] Notwithstanding the un-availability of a Feynman formulation, an abstract version of the

---

[10] A heuristic "sum–over–histories" representation for the dynamics of the electron quantum field (patently distinct from a Feynman "sum-over-histories") was given by W. Tobocman, *Nuovo Cimento* **3**, 1213 (1956).

iteration solution (4.48), (4.51) obtains for a state vector $\boldsymbol{\Psi}$ satisfying (4.70) with $\mathbf{H} = \mathbf{H}_0 + \mathbf{H}_{\text{int}}$ and provides the basis for perturbation-theoretic calculations involving such a quantum field.

3. Neologized Dirac Formulation

By evoking the dynamical equation (4.43), the generic expectation value equation (4.54) can be recast in the form

$$\langle f \rangle = \int \Psi[\phi; 0]^* \mathbf{f} \Psi[\phi; 0] \mathscr{D}(\phi) \tag{4.74}$$

with $\mathbf{H}$ defined by (4.45) and the Hermitian *observable*

$$\mathbf{f} \equiv (\exp it\mathbf{H}/\hbar) f[\phi(x), -i\hbar \, \delta/\delta\phi(x)](\exp -it\mathbf{H}/\hbar) = f[\boldsymbol{\phi}, \boldsymbol{\pi}], \tag{4.75}$$

$$\boldsymbol{\phi} = \boldsymbol{\phi}(x; t) \equiv (\exp it\mathbf{H}/\hbar)\phi(x)(\exp -it\mathbf{H}/\hbar) \tag{4.76}$$

$$\boldsymbol{\pi} = \boldsymbol{\pi}(x; t) \equiv (\exp it\mathbf{H}/\hbar)(-i\hbar \, \delta/\delta\phi(x))(\exp -it\mathbf{H}/\hbar), \tag{4.77}$$

embodying the quantum field dynamics, and the wave functional in (4.74) fixed at the initial instant of time $t = 0$. Equations (4.74) and (4.75) are mathematical statements for the *Heisenberg picture* of quantum field theory with the time rate of change of any expectation value (4.74)

$$\langle \dot{f} \rangle = \int \Psi[\phi; 0]^* \dot{\mathbf{f}} \Psi[\phi; 0] \mathscr{D}(\phi) \tag{4.78}$$

obtained by differentiating (4.75),

$$df/dt \equiv \dot{\mathbf{f}} = [\mathbf{f}, \mathbf{H}], \tag{4.79}$$

where the *Dirac bracket* of two observable operators is defined as in quantum mechanics,

$$[\mathbf{f}, \mathbf{g}] \equiv (i\hbar)^{-1}(\mathbf{fg} - \mathbf{gf}). \tag{4.80}$$

Thus, for example, the components of (4.76) and (4.77) have the Dirac brackets

$$[\boldsymbol{\phi}_i(x; t), \boldsymbol{\phi}_j(y; t)] = [\boldsymbol{\pi}_i(x; t), \boldsymbol{\pi}_j(y; t)] = 0,$$

$$[\boldsymbol{\phi}_i(x; t), \boldsymbol{\pi}_j(y; t)] = \delta_{ij} \, \delta(x - y), \tag{4.81}$$

with the identity operator understood to appear on the right-hand side of the latter equation. In quantum field theory the observable constants of the motion are quantities which have a zero Dirac bracket with the quantum Hamiltonian (4.45).

The set of all Hermitian observables is closed under operator addition and Dirac bracket combination, in the sense that both the sum and the Dirac bracket of any two observables are observables. Hence, the set of all Hermitian observables subject to operator addition and Dirac bracket combination constitutes a Lie algebra. Moreover, any subset of Hermitian observables that is closed with respect to operator addition and Dirac bracket combination constitutes a Lie subalgebra of the Lie algebra of all Hermitian observables. Equations (3.28) and (4.79) for the time rate of change of observables in field theory are identical except for the meaning of the bracket combination law, according to Poisson in classical field theory with (3.29) and according to Dirac in quantum field theory with (4.80). A Lie algebraic structure for the set of all observables is provided in either case by the bracket combination law, but the entire Lie algebraic structure for classical fields with (3.29) cannot be preserved for quantum fields with (4.80) by the *Dirac correspondence*

$$[f, g] = h \Rightarrow [\mathbf{f}, \mathbf{g}] = \mathbf{h}, \tag{4.82}$$

even though extra special properties of the Poisson bracket for fields hold good with proper ordering under the Dirac correspondence. As for dynamical systems with a finite number of degrees of freedom, only a certain elite subset of the field-theoretic observables follows the Dirac correspondence (4.82), in the sense that for any triplet of observables f, g, h in the subset with $[f, g] = h$ the triplet of associated Hermitian observables $\mathbf{f}, \mathbf{g}, \mathbf{h}$ is such that $[\mathbf{f}, \mathbf{g}] = \mathbf{h}$. Because the Dirac correspondence carries Eqs. (3.40) into Eqs. (4.81) for components of the canonical field variable n-tuples, the elements of the infinite-dimensional Lie algebra $\mathfrak{A}_s$ [observables of the form (3.30)] follow the Dirac correspondence with the unique Hermitian observable

$$\mathbf{f} = \tfrac{1}{2} \iint (\boldsymbol{\phi}(x; t) \cdot \alpha(x, y) \cdot \boldsymbol{\phi}(y; t) + \boldsymbol{\phi}(x; t) \cdot \beta(x, y) \cdot \boldsymbol{\pi}(y; t)$$

$$+ \, \boldsymbol{\pi}(x; t) \cdot \beta'(x, y) \cdot \boldsymbol{\phi}(y; t) + \boldsymbol{\pi}(x; t) \cdot \gamma(x, y) \cdot \boldsymbol{\pi}(y; t)) \, dx \, dy$$

$$+ \int (\xi(z) \cdot \boldsymbol{\phi}(z; t) + \eta(z) \cdot \boldsymbol{\pi}(z; t)) \, dz + \omega \tag{4.83}$$

associated with the generic element (3.30), where $\beta'_{ij}(x, y) \equiv \beta_{ji}(y, x)$ in (4.83). The subset of Hermitian observables having the generic form (4.83) is closed with respect to Dirac bracket combination, and hence constitutes an infinite-dimensional Lie algebra $\mathfrak{A}_s$, the quantum correspondent of $\mathfrak{A}_s$. Since the Lie algebraic structure of $\mathfrak{A}_s$ with the Poisson bracket combination law is identical to the Lie algebraic structure of $\mathfrak{A}_s$ with the Dirac bracket combination law, the Lie algebras $\mathfrak{A}_s$ and $\mathfrak{A}_s$ are isomorphic. However, in analogy to the situation that prevails in quantum mechanics, observables with higher order ϕ, π product terms cannot be assigned an Hermitian ordering which validates the Dirac correspondence (4.82) for them and other observables.

To the Lie algebra $\mathscr{A}$ of a subset of field-theoretic Hermitian observables that is closed with respect to operator addition and Dirac bracket combination, there is associated an infinite-dimensional conjugate Lie algebra $\mathscr{A}^*$ consisting of all observables which have a zero Dirac bracket with all elements of $\mathscr{A}$. As in quantum mechanics, the observable constants of the motion constitute the Lie algebra $\mathscr{A}_H^* \equiv \{\mathbf{f} : [\mathbf{f}, \mathbf{H}] = 0\}$ conjugate to the one-dimensional Lie algebra consisting of elements proportional to the quantum Hamiltonian, $\mathscr{A}_H \equiv \{(\text{real constant parameter}) \times \mathbf{H}\}$. Associated with the infinite-dimensional Lie algebra $\mathscr{A}_H^*$, we have the Lie algebra $\hat{\mathscr{A}}_H^*$ consisting of a maximal set, closed under Dirac bracket combination, of linearly independent constants of the motion (exclusive of $\mathbf{H}$ and the trivial generic observables that are constants of the motion for any $\mathbf{H}$). If the action $S = S[\phi(x; t)]$ is a general quadratic expression in $\phi(x; t)$, that is, if the Lagrangian is of the form (4.16), the operator Euler–Lagrange field equations (4.26) are equivalent to inhomogeneous linear equations in the operator field amplitude n-tuple (4.76), namely

$$\int (\rho(x, y) \cdot \ddot{\boldsymbol{\phi}}(y; t) + \theta(x, y) \cdot \dot{\boldsymbol{\phi}}(y; t) - \zeta(x, y) \cdot \boldsymbol{\phi}(y; t)) \, dy + \xi(x) = 0,$$

$$(4.84)$$

and for such a generic linear field theory $\hat{\mathscr{A}}_H^*$ is an infinite-dimensional Lie algebra; it is likely that $\hat{\mathscr{A}}_H^*$ is also infinite-dimensional for a generic nonlinear field theory. In general, the classical and quantum Lie algebras $\mathscr{A}_H^*$ and $\mathscr{A}_H^*$ are not isomorphic for there occur elements of $\mathscr{A}_H^*$ with ϕ, π product terms of higher order than bilinear in the canonical variables. However, $\hat{\mathscr{A}}_H^*$ and $\hat{\mathscr{A}}_H^*$ are generally isomorphic for a linear field theory, as exemplified by the field theory associated with the simple Hamiltonian (3.32) for which $\hat{\mathscr{A}}_H^*$ consists of all real linear combinations of the Hermitian observable constants of the motion

$$\mathbf{f}_i = \int (\boldsymbol{\pi} \cdot (-\nabla^2)^i \boldsymbol{\pi} + \boldsymbol{\phi} \cdot (-\nabla^2)^{i+1} \boldsymbol{\phi}) \, dx \qquad (i = \pm 1, \pm 2, \ldots).$$

$$(4.85)$$

It is an open question whether it is possible to prescribe Hermitian orderings which take each element of $\mathscr{A}_H{}^*$ into an element of $\mathscr{A}_\mathbf{H}{}^*$ in such a manner that $\mathscr{A}_H{}^*$ and $\mathscr{A}_\mathbf{H}{}^*$ are isomorphic in the case of a generic nonlinear field theory.

A transformation of all observables

$$\mathbf{f} = f[\boldsymbol{\phi}, \boldsymbol{\pi}] \overset{w}{\to} \mathbf{f}_\tau = f_\tau[\boldsymbol{\phi}, \boldsymbol{\pi}]$$

$$\mathbf{g} = g[\boldsymbol{\phi}, \boldsymbol{\pi}] \overset{w}{\to} \mathbf{g}_\tau = g_\tau[\boldsymbol{\phi}, \boldsymbol{\pi}]$$

$$\cdots$$

$$(4.86)$$

such that

$$\mathbf{f}_0 = \mathbf{f}, \qquad \mathbf{g}_0 = \mathbf{g}, \qquad \cdots \qquad (4.87)$$

is said to be a *canonical transformation* generated by the Hermitian observable $\mathbf{w} = w[\phi(x), -i\hbar \, \delta/\delta\phi(x)]$, considered here to be independent of t, if the transformed observables satisfy the first-order total differential equations

$$d\mathbf{f}_\tau/d\tau = [\mathbf{f}_\tau, \mathbf{w}], \qquad d\mathbf{g}_\tau/d\tau = [\mathbf{g}_\tau, \mathbf{w}], \qquad \cdots. \qquad (4.88)$$

A transformed observable is given explicitly in terms of the original observable by the Maclaurin series derived from (4.87) and (4.88),

$$\mathbf{f}_\tau = \mathbf{f} + \tau[\mathbf{f}, \mathbf{w}] + (\tau^2/2!)[[\mathbf{f}, \mathbf{w}], \mathbf{w}] + \cdots$$

$$= (\exp i\tau\mathbf{w}/\hbar)\mathbf{f}(\exp - i\tau\mathbf{w}/\hbar). \qquad (4.89)$$

As for canonical transformations in quantum mechanics, we have the preservation of Dirac bracket relations by canonical transformations in quantum field theory,

$$\mathbf{h} = [\mathbf{f}, \mathbf{g}] \Rightarrow \mathbf{h}_\tau = [\mathbf{f}_\tau, \mathbf{g}_\tau]. \qquad (4.90)$$

Thus, Lie algebras of Hermitian observables are mapped into isomorphic Lie algebras by canonical transformations. Again as in quantum mechanics, the time rate of change of any canonically transformed observable is prescribed by a dynamical equation of the form (4.79),

$$d\mathbf{f}_\tau/dt \equiv \dot{\mathbf{f}}_\tau = [\mathbf{f}_\tau, \mathbf{H}_\tau], \qquad (4.91)$$

as seen by putting $\mathbf{h} = \mathbf{f}$ and $\mathbf{g} = \mathbf{H}$ in (4.90). By specializing (4.90) to the cases for which $\mathbf{h} = c$, a quantity independent of $\boldsymbol{\phi}$ and $\boldsymbol{\pi}$, we find that

$$[\mathbf{f}, \mathbf{g}] = c \Rightarrow [\mathbf{f}_\tau, \mathbf{g}_\tau] = c. \tag{4.92}$$

In particular, for the components of the field amplitude and momentum density n-tuples, the Dirac bracket Eqs. (4.81) yield

$$[\boldsymbol{\phi}_{\tau i}(x;t), \boldsymbol{\phi}_{\tau j}(y;t)] = [\boldsymbol{\pi}_{\tau i}(x;t), \boldsymbol{\pi}_{\tau j}(y;t)] = 0, \tag{4.93}$$

$$[\boldsymbol{\phi}_{\tau i}(x;t), \boldsymbol{\pi}_{\tau j}(y;t)] = \delta_{ij}\,\delta(x - y). \tag{4.94}$$

Specializing to the cases for which $\mathbf{f}_\tau$ equals each of the $2n$ components of $\boldsymbol{\phi}_\tau(x)$ and $\boldsymbol{\pi}_\tau(x)$ in (4.91), we immediately obtain dynamical equations for the transformed canonical field variables,

$$\dot{\boldsymbol{\phi}}_\tau(x;t) = [\boldsymbol{\phi}_\tau(x;t), \mathbf{H}_\tau], \qquad \dot{\boldsymbol{\pi}}_\tau(x;t) = [\boldsymbol{\pi}_\tau(x;t), \mathbf{H}_\tau]. \tag{4.95}$$

In quantum field theory one defines the transformed wave functional at the initial instant of time as

$$\Psi_\tau[\phi; 0] \equiv (\exp i\tau \mathbf{w}/\hbar)\Psi[\phi; 0], \tag{4.96}$$

so that expectation values (4.74) are invariant with respect to canonical transformations,

$$\int \Psi_\tau[\phi; 0]^* \mathbf{f}_\tau \Psi_\tau[\phi; 0]\mathcal{D}(\phi) = \langle f\rangle. \tag{4.97}$$

Hence, canonical transformations in quantum field theory have no effect on observable predictions.

For canonical transformations generated by an m-dimensional Lie algebra $\mathscr{A}$ composed of all real linear combinations of m linearly independent Hermitian observables $\mathbf{w}_1, \ldots, \mathbf{w}_m$ with each $\mathbf{w}_i = w_i[\phi(x), -i\hbar\,\delta/\delta\phi(x)]$ independent of time, we have

$$[\mathbf{w}_i, \mathbf{w}_j] = -\sum_{k=1}^{m} c_{ijk}\,\mathbf{w}_k \tag{4.98}$$

with the structure constants $c_{ijk} = -c_{jik}$ satisfying the quadratic Lie identities (1.52); the associated generators for canonical transformations

$$\mathbf{G}_i \equiv (i/\hbar)\mathbf{w}_i = (i/\hbar)w_i[\phi(x), -i\hbar\,\delta/\delta\phi(x)] \tag{4.99}$$

are m functional differential operators that satisfy the commutation relations

$$\mathbf{G}_i\,\mathbf{G}_j - \mathbf{G}_j\,\mathbf{G}_i = \sum_{k=1}^{m} c_{ijk}\,\mathbf{G}_k. \tag{4.100}$$

As for systems with a finite number of degrees of freedom, the classical generators (3.46) differ strikingly from the quantum generators (4.99) for canonical transformations in field theory, the latter generators exhibiting a general nonlinear dependence on $\delta/\delta\phi(x)$ with $\pi(x)$ and $\delta/\delta\pi(x)$ absent. A linear representation of the m-parameter Lie group associated with the Lie algebra $\mathscr{A}$ is provided by linear differential operators of the form

$$(\exp \alpha \cdot \mathbf{G}) \equiv \sum_{N=0}^{\infty} (N!)^{-1}(\alpha \cdot \mathbf{G})^N,$$

parametrized by a real m-tuple $\alpha = (\alpha_1, \ldots, \alpha_m)$. Under canonical transformations generated by $\mathscr{A}$, a generic Hermitian observable (4.75) is transformed to

$$\mathbf{f}_\alpha = (\exp \alpha \cdot \mathbf{G})\mathbf{f}(\exp - \alpha \cdot \mathbf{G}) \tag{4.101}$$

with all properties derived for the transformed observables (4.89) featured by the transformed observables (4.101), and the transformed wave functional at the initial instant of time defined as

$$\Psi_\alpha[\phi; 0] \equiv (\exp \alpha \cdot \mathbf{G})\Psi[\phi; 0]. \tag{4.102}$$

As an illustration of quantum canonical transformations generated by a finite-dimensional Lie algebra, consider the three-dimensional subalgebra of $\mathfrak{A}_s$,

$$\mathscr{A} = \Big\{ \alpha_1 \mathbf{w}_1 + \alpha_2 \mathbf{w}_2 + \alpha_3 \mathbf{w}_3 :$$

$$\mathbf{w}_1 \equiv -\frac{1}{4}\int \left(\hbar^2\,\frac{\delta}{\delta\phi(x)} \cdot \frac{\delta}{\delta\phi(x)} + \phi(x) \cdot \phi(x) \right) dx,$$

$$\mathbf{w}_2 \equiv \frac{1}{4}\int \left(\hbar^2\,\frac{\delta}{\delta\phi(x)} \cdot \frac{\delta}{\delta\phi(x)} - \phi(x) \cdot \phi(x) \right) dx,$$

$$\mathbf{w}_3 \equiv \frac{i\hbar}{4}\int \left(\phi(x) \cdot \frac{\delta}{\delta\phi(x)} + \frac{\delta}{\delta\phi(x)} \cdot \phi(x) \right) dx \Big\}, \tag{4.103}$$

which is the quantum correspondent of (3.48). For this $\mathscr{A}$ the Dirac bracket closure relations (4.98) are

$$[\mathbf{w}_1, \mathbf{w}_2] = -\mathbf{w}_3, \qquad [\mathbf{w}_2, \mathbf{w}_3] = -\mathbf{w}_1, \qquad [\mathbf{w}_3, \mathbf{w}_1] = \mathbf{w}_2, \qquad (4.104)$$

and the generators (4.99),

$$\mathbf{G}_1 = -\frac{i}{4} \int \left(\hbar \frac{\delta}{\delta\phi(x)} \cdot \frac{\delta}{\delta\phi(x)} + \hbar^{-1}\phi(x) \cdot \phi(x) \right) dx$$

$$\mathbf{G}_2 = \frac{i}{4} \int \left(\hbar \frac{\delta}{\delta\phi(x)} \cdot \frac{\delta}{\delta\phi(x)} - \hbar^{-1}\phi(x) \cdot \phi(x) \right) dx \qquad (4.105)$$

$$\mathbf{G}_3 = -\frac{1}{4} \int \left(\phi(x) \cdot \frac{\delta}{\delta\phi(x)} + \frac{\delta}{\delta\phi(x)} \cdot \phi(x) \right) dx,$$

satisfy the commutation relations (4.100) with the independent nonzero structure constants

$$c_{123} = c_{231} = -c_{312} = 1.$$

Because the Lie algebra (4.103) is a subalgebra of $\mathfrak{A}_s$, it is isomorphic to the corresponding subalgebra of $\mathfrak{A}_s$, namely (3.48), and thus $\mathscr{A}$ defined by (4.103) is isomorphic to the Lie algebra for $SL(2, R)$. The generators (4.105) give a representation of $SL(2, R)$ canonical transformations on the space of wave functionals according to (4.102).

The dynamical evolution of all observables, prescribed by (4.79), can be viewed as a canonical transformation parametrized by time and generated by the quantum Hamiltonian, as seen by putting $\tau = t$, $\mathbf{w} = \mathbf{H} = H[\phi(x), -i\hbar \, \delta/\delta\phi(x)]$, and $\mathbf{f}_t \equiv f_t[\phi(x), -i\hbar \, \delta/\delta\phi(x)] \equiv f[\phi(x; t), \pi(x; t)]$ in (4.88). Under this special canonical transformation, we have $\phi(x; 0) = \phi(x) \overset{\mathbf{H}}{\to} \phi(x; t)$ and $\pi(x; 0) = (-i\hbar \, \delta/\delta\phi(x)) \overset{\mathbf{H}}{\to} \pi(x; t)$, the canonical field variable operators defined by (4.76) and (4.77).

Since the Lie algebra $\mathscr{A}_{\mathbf{H}}{}^*$ is infinite-dimensional for a generic linear field theory and conjectured to be infinite-dimensional for a generic nonlinear field theory, the notion of canonical transformations generated by an m-dimensional Lie algebra must be extended to the case of an infinite-dimensional Lie algebra in order to apply to an $\mathscr{A}_{\mathbf{H}}{}^*$ in quantum field theory. We have $\mathbf{G}_i\mathbf{H} = \mathbf{H}\mathbf{G}_i$ for the generators associated with $\mathscr{A}_{\mathbf{H}}{}^*$, and thus the Hamiltonian is invariant for canonical transformations generated by the Lie algebra $\mathscr{A}_{\mathbf{H}}{}^*$, $\mathbf{H}_\alpha \equiv \mathbf{H}$, with α such that the sum in (4.101), $\alpha \cdot \mathbf{G} \equiv \sum_{i=1}^{\infty} \alpha_i \mathbf{G}_i$, is convergent. Because the dynamics of Hermitian observables is unchanged by these canonical transformations, the Lie algebra $\mathscr{A}_H{}^*$ is referred to as the *symmetry algebra* for the quantum field theory, and the associated Lie group $\mathscr{G}_H{}^*$, with its linear representation on the space of wave functionals $\mathscr{G}_H{}^* = \{(\exp \alpha \cdot \mathbf{G})\}$, is called the *symmetry group*.

Finally, we note the especially simple kind of dynamics that arises if H is an element of the Lie algebra $\mathfrak{A}_s$ discussed above, that is, if H has the form of the right-hand side of (3.30) and $\mathbf{H}$ the right-hand side of (4.83), as in the case of a generic linear field theory with operator Euler–Lagrange field equations of the form (4.84). Because of the isomorphism of $\mathfrak{A}_s$ to $\mathfrak{A}_s$, the quantum dynamics associated with an $\mathbf{H}$ in $\mathfrak{A}_s$ is identical to the classical dynamics associated with the corresponding H in $\mathfrak{A}_s$ for corresponding observables in $\mathfrak{A}_s$ and $\mathfrak{A}_s$. In particular, the components of $\phi(x; t)$ are inhomogeneous linear expressions in the components of $\phi(x; 0) = \phi(x)$ and $\pi(x; 0) = -i\hbar \, \delta/\delta\phi(x)$ with the same time-dependent coefficients as in the classical theory, and likewise for the components of $\pi(x; t)$. Dynamical containment in $\mathfrak{A}_s$ is featured by all elements in the Lie algebra if $\mathbf{H}$ is in $\mathfrak{A}_s$,

$$f[\phi(x; t), \pi(x; t)]$$

$$\equiv f_t[\phi(x; 0), \pi(x; 0)]$$

$$= \tfrac{1}{2} \iint (\phi(x; 0) \cdot \alpha(x, y; t) \cdot \phi(y; 0) + \phi(x; 0) \cdot \beta(x, y; t) \cdot \pi(y; 0)$$

$$+ \pi(x; 0) \cdot \beta'(x, y; t) \cdot \phi(y; 0)$$

$$+ \pi(x; 0) \cdot \gamma(x, y; t) \cdot \pi(y; 0)) \, dx \, dy + \int (\xi(z; t) \cdot \phi(z; 0)$$

$$+ \eta(z; t) \cdot \pi(z; 0)) \, dz + \omega(t), \tag{4.106}$$

where $\beta'_{ij}(x, y; t) \equiv \beta_{ji}(y, x; t)$ and with $\alpha(x, y; t)$, $\beta(x, y; t)$, $\gamma(x, y; t)$, $\xi(z; t)$, $\eta(z; t)$, and $\omega(t)$ the same as in (3.51) because of (4.79), the Lie algebra addition and Dirac bracket closure properties, and the isomorphism of $\mathfrak{A}_s$ to $\mathfrak{A}_s$. A generic linear field theory has solvable dynamics on the quantum as well as on the classical level because $\mathfrak{A}_s$ is isomorphic to $\mathfrak{A}_s$.

PROSPECTS

Notwithstanding the elegance and simplicity of its present form, the conceptual framework of quantum field theory is likely to undergo dramatic changes in future years if quantum field theory is not superseded entirely by a new and different dynamical theory for elementary particles. In quantum electrodynamics the technical procedure called "renormalization" is required to derive finite predictions for experimentally observable quantities from the mathematical formulations. The status of quantum electrodynamics with renormalization has been described by Heitler and by Jauch and Rohrlich:

> The ambiguities can always be settled by applying a certain amount of "wishful mathematics," namely, by using additional conditions for the evaluation of ambiguous integrals. Unless such conditions are used, the results of the theory may contradict its very foundation.... Clearly such a mathematical situation is unacceptable.
>
> On the other hand, these difficulties do not prevent us from giving a theoretical answer to every legitimate question concerning observable effects. These answers are, wherever they can be tested, always in excellent agreement with the facts, and no serious discrepancy exceeding the limits of accuracy of the calculation has so far been discovered.[1]

With respect to the applications, i.e., the actual description of physical quantities, we have here one of the best-established physical theories. Whenever the theory is subjected to an experimental test, we find the theoretical prediction in complete agreement with the experimental result. The accuracy

[1] W. Heitler, "The Quantum Theory of Radiation," p. 354. Oxford Univ. Press (Clarendon), London and New York, 1954.

93

in the agreement is limited only by the experimental error and the endurance and ingenuity of the computer.

With respect to the fundamental concepts, on the contrary, we are not so fortunate. The theory is incomplete insofar as we are forced to introduce the charge of the electron and the masses of electron and photon as phenomenological quantities. The value of these quantities cannot be deduced from the theory but must be accepted as empirically given. This point of view enabled us to carry through the program of renormalization which was essential for the removal of the divergences in the iteration solution.[2]

Thus, the renormalization procedure in quantum electrodynamics can at best be accepted as tentative and heuristic. The alternative negative attitude, outright rejection of renormalized quantum electrodynamics, is in fact advocated by Dirac:

It seems to be quite impossible to put this theory on a mathematically sound basis. At one time physical theory was all built on mathematics that was inherently sound. I do not say that physicists always use sound mathematics; they often use unsound steps in their calculations. But previously when they did so it was simply because of, one might say, laziness. They wanted to get results as quickly as possible without doing unnecessary work. It was always possible for the pure mathematician to come along and make the theory sound by bringing in further steps, and perhaps by introducing quite a lot of cumbersome notation and other things that are desirable from a mathematical point of view in order to get everything expressed rigorously but do not contribute to the physical ideas. The earlier mathematics could always be made sound in that way, but in the renormalization theory we have a theory that has defied all the attempts of the mathematician to make it sound. I am inclined to suspect that the renormalization theory is something that will not survive in the future, and that the remarkable agreement between its results and experiment should be looked on as a fluke.[3]

If the renormalization procedure in quantum electrodynamics is in essence illegitimate, the entire conceptual framework of quantum field theory must be suspect. With regard to this possibility, we recall the opinion of extreme dissent by Einstein:

[2] J. M. Jauch and F. Rohrlich, "The Theory of Photons and Electrons," pp. 415–416. Addison–Wesley, Cambridge, Massachusetts, 1955. Reprinted by permission.

[3] P. A. M. Dirac, *Sci. Amer.* **208**, 50 (May 1963).

Is it conceivable that a classical field theory permits one to understand the atomistic and quantum structure of reality? Almost everybody will answer this question with "no." But I believe that at the present time nobody knows anything reliable about it. This is so because we cannot judge in what manner and how strongly the exclusion of singularities reduces the manifold of solutions. We do not possess any method at all to derive systematically solutions that are free of singularities. Approximation methods are of no avail since one never knows whether or not there exists to a particular appoximate solution an exact solution free of singularities. For this reason we cannot at present compare the content of a nonlinear classical field theory with experience. Only a significant progress in the mathematical methods can help here. At the present time the opinion prevails that a field theory must first, by "quantization," be transformed into a statistical theory of field probabilities according to more or less established rules. I see in this method only an attempt to describe relationships of an essentially nonlinear character by linear methods.[4]

[4] A. Einstein, "The Meaning of Relativity," p. 165. Princeton Univ. Press, Princeton, New Jersey, 1966. Reprinted by permission.

Appendix A *Functional Differentiation*

Let the ordered array of Cartesian coordinates $x = (x_1, \ldots, x_m)$ represent a point in the m-dimensional Euclidean space R_m, and let $f(x) = (f_1(x), \ldots, f_n(x))$ denote a real n-tuple function of x. A *functional* $F = F[f(x)]$, a rule which assigns a number F to a function $f(x)$, is *continuous* at $f(x)$ if [1]

$$\lim_{\varepsilon \to 0} F[f(x) + \varepsilon\sigma(x)] = F[f(x)] \qquad (A.1)$$

for all real n-tuple functions of x, $\sigma(x) = (\sigma_1(x), \ldots, \sigma_n(x))$, in a prescribed space $\mathscr{F} = \{\sigma(x)\}$. Quite generally, an appropriate function space $\mathscr{F}$ for a well-defined definition of continuity (A.1) is such that the $f(x)$ domain of $F[f(x)]$ and $F[f(x) + \varepsilon\sigma(x)]$ coincide for all $\sigma(x)$ in $\mathscr{F}$; for example, $\mathscr{F}$ might contain all functions of continuity class C^N (perhaps also of "compact support," $\sigma(x)$ equal to zero outside a certain compact region associated with each function) that vanish at prescribed boundary values of x. While the domain of a functional is in general a *class* of functions, the sum of two functions in the domain not

---

[1] For the original versions of functional continuity and functional differentiation, see V. Volterra, "Theory of Functional and of Integral and Integro-Differential Equations." Dover, New York, 1959.

necessarily being in the domain, $\mathscr{F}$ is patently a *space* of functions, closed under addition.[2] For example, the functional

$$F[f(x)] = \int_0^1 \cdots \int_0^1 f(x) \cdot f(x)\, dx \equiv \int_0^1 \cdots \int_0^1 \left(\sum_{i=1}^n (f_i(x))^2 \right) \left(\prod_{j=1}^m dx_j \right)$$

(A.2)

is continuous for all continuous $f(x)$ with $\mathscr{F} \equiv \{\sigma(x) : \sigma(x) \in C^0\}$. A functional $F = F[f(x)]$, continuous at $f(x)$, is *differentiable* at $f(x)$ if [1]

$$((d/d\varepsilon)F[f(x) + \varepsilon\sigma(x)])_{\varepsilon=0} \equiv \int_{R_m} \sigma(x) \cdot [\delta F/\delta f(x)]\, dx$$

$$\equiv \int_{R_m} \left(\sum_{i=1}^n \sigma_i(x)[\delta F/\delta f_i(x)] \right) \left(\prod_{j=1}^m dx_j \right) \quad \text{(A.3)}$$

exists as a linear functional in $\sigma(x)$, the latter n-tuple function again being any element in the prescribed $\mathscr{F}$. From the implicit definition (A.3) it follows immediately that functional differentiation is a linear operation,

$$[\delta/\delta f(x)]\,(c_1 F_1 + c_2 F_2) \equiv c_1[\delta F_1/\delta f(x)] + c_2[\delta F_2/\delta f(x)] \quad \text{(A.4)}$$

that obeys the differentiation product rule,

$$[\delta/\delta f(x)]\,(F_1 F_2) \equiv [\delta F_1/\delta f(x)]\,F_2 + F_1\,[\delta F_2/\delta f(x)]. \quad \text{(A.5)}$$

In fact, for a differentiable function of a functional, $\phi = \phi(F) = \phi(F[f(x)])$, we have

$$\delta\phi/\delta f(x) = (d\phi/dF)\,[\delta F/\delta f(x)]. \quad \text{(A.6)}$$

Illustrations of functional derivatives calculated according to (A.3) are provided by the following formulas:

$$\frac{\delta}{\delta f_i(x)} \int_\Omega f(x') \cdot f(x')\, dx' = \begin{array}{ll} 2f_i(x) & \text{for } x \in \Omega \\ 0 & \text{for } x \notin \Omega, \end{array} \quad \text{(A.7)}$$

$$\Omega \text{ compact in } R_m, \qquad \mathscr{F} \equiv \{\sigma(x): \sigma(x) \in C^0\},$$

---

[2] A comprehensive survey of special spaces $\mathscr{F}$ for *distributions* [specialized *linear functionals* with $F[f(x) + \varepsilon\sigma(x)] = F[f(x)] + \varepsilon F[\sigma(x)]$ for all $\sigma(x)$ in $\mathscr{F}$] appears in L. Hörmander, "Linear Partial Differential Operators," pp. 33–62. Academic Press, New York, 1964.

$$\delta f_j(a)/\delta f_i(x) = \delta_{ij}\,\delta(x - a) \equiv \delta_{ij}\prod_{k=1}^{m}\delta(x_k - a_k),$$

$$\mathscr{F} \equiv \{\sigma(x): \sigma(x) \in C^0\}, \tag{A.8}$$

$$\frac{\delta}{\delta f_i(x)}\int_{\Omega}\sum_{j=1}^{m}\frac{\partial f(x')}{\partial x_j'}\cdot\frac{\partial f(x')}{\partial x_j'}\,dx' = -2\sum_{j=1}^{m}\frac{\partial^2 f_i(x)}{\partial x_j\,\partial x_j} \qquad \text{for}\quad x \in \Omega,$$

$$\mathscr{F} \equiv \{\sigma(x): \sigma(x) \in C^1,\ \sigma(x) = 0 \quad \text{for}\quad x \in \partial\Omega\}, \tag{A.9}$$

$$\frac{\delta}{\delta f_i(x)}\left(\exp - c\int_{R_m} f(x')\cdot f(x')\,dx'\right)$$

$$= -2cf_i(x)\left(\exp - c\int_{R_m} f(x')\cdot f(x')\,dx'\right), \tag{A.10}$$

$$\mathscr{F} \equiv \left\{\sigma(x): \int_{R_m}\sigma(x)\cdot\sigma(x)\,dx < \infty\right\}.$$

Note that it is unnecessary to give a meaning to the so-called variational symbol δ when δ stands alone although the variational form of (1.3), obtained by expanding $F[f(x) + \varepsilon\sigma(x)]$ as a Maclaurin series in ε and setting $\varepsilon\sigma(x) \equiv \delta f(x)$,

$$\delta F[f(x)] \equiv F[f(x) + \delta f(x)] - F[f(x)]$$

$$= \int_{R_m}\delta f(x)\cdot[\delta F/\delta f(x)]\,dx + O(\delta f(x)\cdot\delta f(x)), \tag{A.11}$$

is quite useful as a computational device. A functional $F = F[f(x)]$ has an *extremum* at $f(x)$ if

$$\delta F/\delta f(x) = 0 \tag{A.12}$$

when the left-hand side is evaluated at $f(x)$.

Higher-order functional differentiation is defined by iteration of (A.3). Thus, the second-order functional derivatives of $F = F[f(x)]$ are given implicitly by

$$\left(\frac{d^2}{d\varepsilon^2}F[f(x) + \varepsilon\sigma(x)]\right)_{\varepsilon=0} \equiv \int_{R_m \times R_m}\sigma(x)\cdot\frac{\delta^2 F}{\delta f(x)\,\delta f(y)}\cdot\sigma(y)\,dx\,dy$$

$$\equiv \int_{R_m \times R_m}\sum_{i,j=1}^{n}\sigma_i(x)\sigma_j(y)\frac{\delta^2 F}{\delta f_i(x)\,\delta f_j(y)}\,dx\,dy. \tag{A.13}$$

It follows from (A.13) that second-order functional differentiation is symmetrical, linear, and obeys the iterated differentiation product rule,

$$\delta^2 F/\delta f(x)\,\delta f(y) \equiv \delta^2 F/\delta f(y)\,\delta f(x), \tag{A.14}$$

$$\frac{\delta^2}{\delta f(x)\,\delta f(y)}(c_1 F_1 + c_2 F_2) \equiv c_1\frac{\delta^2 F_1}{\delta f(x)\,\delta f(y)} + c_2\frac{\delta^2 F_2}{\delta f(x)\,\delta f(y)}, \tag{A.15}$$

$$\frac{\delta^2}{\delta f(x)\,\delta f(y)}(F_1 F_2) \equiv \frac{\delta^2 F_1}{\delta f(x)\,\delta f(y)}F_2 + \frac{\delta F_1}{\delta f(x)}\frac{\delta F_2}{\delta f(y)}$$

$$+ \frac{\delta F_1}{\delta f(y)}\frac{\delta F_2}{\delta f(x)} + F_1\frac{\delta^2 F_2}{\delta f(x)\,\delta f(y)}. \tag{A.16}$$

The variational form of (A.13), obtained by making use of (A.11), facilitates actual computation of second-order functional derivatives,

$$\delta\left(\frac{\delta F}{\delta f_i(x)}\right) = \int_{R_m}\sum_{j=1}^{n}\frac{\delta^2 F}{\delta f_i(x)\,\delta f_j(y)}\,\delta f_j(y)\,dy + O(\delta f(y)\cdot\delta f(y)). \tag{A.17}$$

The most important second-order functional differential operator is the *functional Laplacian*

$$\int_{R_m}\frac{\delta}{\delta f(x)}\cdot\frac{\delta}{\delta f(x)}\,dx \equiv \int_{R_m}\left(\lim_{y\to x}\sum_{i=1}^{n}\frac{\delta^2}{\delta f_i(x)\,\delta f_i(y)}\right)dx, \tag{A.18}$$

which plays a key role in quantum field theory.

In analogy to the Taylor expansion of an analytic function of several real variables, by definition, an *analytic functional* $F = F[f(x)]$ admits a *Volterra expansion*[1] about a prescribed function $\bar{f} = \bar{f}(x)$,

$$F[f(x)] = F[\bar{f}(x)] + \int_{R_m}\frac{\delta F}{\delta f(x)}\bigg|_{f=\bar{f}}\cdot(f(x) - \bar{f}(x))\,dx$$

$$+ \frac{1}{2!}\int_{R_m\times R_m}(f(x) - \bar{f}(x))\cdot\frac{\delta^2 F}{\delta f(x)\,\delta f(y)}\bigg|_{f=\bar{f}}$$

$$\cdot(f(y) - \bar{f}(y))\,dx\,dy + \cdots. \tag{A.19}$$

Other basic notions from the theory of a function of several real variables with partial differentiation carry over to the theory of functionals with functional differentiation. Thus, for example, a homogeneous functional of order α, $F = F[f(x)] \equiv \lambda^{-\alpha}F[\lambda f(x)]$ for all real

$\lambda > 0$, satisfies the linear functional differential equation first noted by Volterra,[1,3]

$$\int_{R_m} f(x) \cdot [\delta F/\delta f(x)] \, dx = \alpha F, \qquad (A.20)$$

and, conversely, the general integral to (A.20) is given by a homogeneous functional of order α. The manifest analogy with the theory of a function of several real variables can also be exploited to establish other general integrals to linear functional differential equations; for example, we have

$$\delta F/\delta f(x) + f(x)F = 0 \qquad \text{for all} \quad x \in R_m \qquad (A.21)$$

admitting the general integral

$$F = (\text{const}) \exp - \tfrac{1}{2} \int_{R_m} f(x) \cdot f(x) \, dx. \qquad (A.22)$$

[3] Equation (A.20) is obtained most readily by applying the general parameter-differentiation formula

$$(d/d\lambda) \, F[f(x; \lambda)] = \int_{R_m} [\delta F/\delta f(x; \lambda)] \cdot [\partial f(x; \lambda)/\partial \lambda] \, dx \qquad (A.23)$$

that follows from (A.3) and

$$F[f(x; \lambda + \varepsilon)] = F[f(x; \lambda) + \varepsilon \, \partial f(x; \lambda)/\partial \lambda] + O(\varepsilon^2)$$

Appendix B   *Linear Representations of a Lie Group*

Consider a set $\mathscr{G} = \{T(\alpha)\}$ of linear operators[1] $T(\alpha)$, labeled by m real essential parameters, the components of the m-tuple $\alpha = (\alpha_1, \ldots, \alpha_m)$. $\mathscr{G}$ is a *linear representation* of an m-parameter Lie group if[2]:

1. $T(\alpha)$ is an analytic function of α about $\alpha = 0$ with $T(0) = \mathbf{1}$, the identity operator; thus,

$$T(\alpha) = \mathbf{1} + \alpha \cdot G \quad \text{plus terms quadratic and of higher order in } \alpha \quad \text{(B.1)}$$

where the *generators* appear as

$$G_i \equiv (\partial T(\alpha)/\partial \alpha_i)_{\alpha = 0}. \quad \text{(B.2)}$$

The m linear operators (B.2) are linearly independent.

2. Each $T(\alpha)$ has a unique inverse $T(\alpha)^{-1} \equiv T(\bar{\alpha}) \in \mathscr{G}$ such that

$$T(\bar{\alpha})T(\alpha) = T(\alpha)T(\bar{\alpha}) = \mathbf{1}. \quad \text{(B.3)}$$

---

[1] See, for example, N. I. Akhiezer and I. M. Glazman, "Theory of Linear Operators in Hilbert Space," pp. 30–39. Ungar, New York, 1961. In physical applications, $T(\alpha)$ is either a square-matrix that acts on a finite-dimensional vector space or a differential operator that acts on an infinite-dimensional function space.
[2] For the sake of practical transparency, our conditions on the group elements $T(\alpha)$ are somewhat more specific than necessary.

3. For any pair of elements $T(\alpha)$ and $T(\beta)$ in $\mathcal{G}$, their ordered product

$$T(\alpha)T(\beta) = T(\gamma) \tag{B.4}$$

is contained in $\mathcal{G}$ with the m-tuple of parameters

$$\gamma = \gamma(\alpha, \beta) = \alpha + \beta \text{ plus} \tag{B.5}$$

(terms bilinear, quadratic, and of higher order in α and β)

as a consequence of (B.1).

Because of (B.4), we have

$$\frac{\partial T(\alpha)}{\partial \alpha_i} \frac{\partial T(\beta)}{\partial \beta_j} = \sum_{k,l=1}^{m} \frac{\partial \gamma_k}{\partial \alpha_i} \frac{\partial \gamma_l}{\partial \beta_j} \frac{\partial^2 T(\gamma)}{\partial \gamma_k \partial \gamma_l} + \sum_{k=1}^{m} \frac{\partial^2 \gamma_k}{\partial \alpha_i \partial \beta_j} \frac{\partial T(\gamma)}{\partial \gamma_k}, \tag{B.6}$$

so by setting $\alpha = \beta = 0$ and making use of (B.5) and the definition (B.2), we obtain

$$G_i G_j = \left(\frac{\partial^2 T(\gamma)}{\partial \gamma_i \partial \gamma_j}\right)_{\gamma=0} + \sum_{k=1}^{m} \left(\frac{\partial^2 \gamma_k}{\partial \alpha_i \partial \beta_j}\right)_{\alpha=\beta=0} G_k. \tag{B.7}$$

From (B.7) it follows that

$$G_i G_j - G_j G_i = \sum_{k=1}^{m} c_{ijk} G_k. \tag{B.8}$$

where the *structure constants* appear as

$$c_{ijk} \equiv \left(\frac{\partial^2 \gamma_k}{\partial \alpha_i \partial \beta_j} - \frac{\partial^2 \gamma_k}{\partial \alpha_j \partial \beta_i}\right)_{\alpha=\beta=0} \equiv -c_{jik}. \tag{B.9}$$

It is readily seen that the structure constants satisfy the quadratic *Lie identities*,

$$\sum_{h=1}^{m} (c_{ijh} c_{hkl} + c_{jkh} c_{hil} + c_{kih} c_{hjl}) \equiv 0, \tag{B.10}$$

by adding the i, j, k cyclic permutations of the commutator of (B.8) with G_k and evoking the linear independence of the $\dot{G}$'s. Lie's main theorem is that an array of constants $c_{ijk} \equiv -c_{jik}$, $[i, j, k = 1, \ldots, m]$, satisfying (B.10) permits the Eqs. (B.8) to be solved for m linear operators G_i related by (B.2) to a $T(\alpha)$ satisfying Eqs. (B.1), (B.3), and (B.4); furthermore, by evoking a trivial relabeling of the elements of $\mathcal{G}$ with a parameter transformation of the form $\alpha \to \alpha +$ (terms quadratic and of higher order in α), elements of $\mathcal{G}$ can be expressed canonically as

$$T(\alpha) = (\exp \alpha \cdot G) \equiv \sum_{N=0}^{\infty} (N!)^{-1}(\alpha \cdot G)^{N}. \qquad \text{(B.11)}$$

The set of all linear combinations of the generators, $\mathscr{A} = \{\alpha \cdot G\}$ with elements parametrized by α, is closed with respect to ordinary operator addition and operator commutation, and hence constitutes a Lie algebra with the Lie product prescribed as the commutator of a pair of elements. A solution to Eqs. (B.8) for the generators, such that no G_i is the zero operator, yields a *representation* of the Lie algebra.[3] If the structure constants are such that c_{ijk} does not vanish for all values of j and k with i any fixed value, we have the so-called adjoint representation with the generators m-dimensional matrices composed of the elements $(G_i)_{jk} = -c_{ijk}$, as readily seen by writing (B.8) with matrix indices and recalling (B.10). The *associated adjoint representation* gives the generators as the differential operators $G_i = -\sum_{j,k=1}^{m} c_{ijk} x_j \, \partial/\partial x_k$ on the space of infinitely differentiable functions of $x = (x_1, \ldots, x_m)$. Lie's main theorem asserts that a representation of the Lie algebra provides a linear representation of the Lie group $\mathscr{G}$ for an appropriate set of α in R_m, with elements in $\mathscr{G}$ given canonically by (B.11). If the appropriate set of α in R_m is a compact point set, the Lie group is said to be *compact*; otherwise, the Lie group is *noncompact*. To obtain all elements for certain Lie groups, it is necessary to augment the set $\{(\exp \alpha \cdot G)\}$ in an obvious way (see Example 4 below).

Linear representations of Lie groups are illustrated by the following examples:

1. $c_{ijk} = 0$, the m-dimensional Abelian Lie group. According to (B.8), all m generators commute with one another. Of particular interest is the differential operator realization $G_i = \partial/\partial x_i$, for which the set of differential operator group elements (B.11),

$$T(\alpha) = \exp(\alpha \cdot \partial/\partial x), \qquad \text{(B.12)}$$

[3] All Lie algebra representations are isomorphic to the Lie algebra of real m-tuples with the addition

$$\alpha + \beta \equiv (\alpha_1 + \beta_1, \ldots, \alpha_m + \beta_m)$$

and the Lie product

$$[\alpha, \beta]_k \equiv \sum_{i,j=1}^{m} c_{ijk} \alpha_i \beta_j$$

The integrability property (1.30) is satisfied for this Lie product because of (B.10). Notice that for $m = 3$ and $c_{ijk} = \varepsilon_{ijk}$, this Lie product is the 3-tuple cross-product of Cartesian vector analysis.

translates the m-tuple argument of a C^∞ function $f(x)$ by α, $T(\alpha)f(x) = f(x + \alpha)$. In order for all elements in the group to be included, each of the m parameters α_i must assume all finite real values, and hence the m-dimensional Abelian Lie group is noncompact. The special case $m = 1$ is closely associated with solutions of first-order (ordinary or partial) linear differential equations.

2. $m = 2$ with $c_{121} = 0$ and $c_{122} = 1$, the proper affine Lie group on real numbers. It is easy to verify that the quadratic Lie identities (B.10) are satisfied. The generator equations (B.8) produce the single relation $G_1 G_2 - G_2 G_1 = G_2$, which admits the two-dimensional matrix representation

$$G_1 = \begin{pmatrix} 1 & 0 \\ 0 & 0 \end{pmatrix}, \qquad G_2 = \begin{pmatrix} 0 & 1 \\ 0 & 0 \end{pmatrix}. \tag{B.13}$$

By putting this two-dimensional representation into (B.11), we obtain

$$T(\alpha) = \exp \begin{pmatrix} \alpha_1 & \alpha_2 \\ 0 & 0 \end{pmatrix} = \begin{pmatrix} e^{\alpha_1} & \alpha_2(e^{\alpha_1} - 1)/\alpha_1 \\ 0 & 1 \end{pmatrix}. \tag{B.14}$$

The group parameters α_1 and α_2 assume all finite real values, so this Lie group is noncompact. Note that the representation (B.14) acts on the space of vectors of the form $\begin{pmatrix} x \\ 1 \end{pmatrix}$ to produce proper affine transformations of the real number $x \to [e^{\alpha_1}x + \alpha_2(e^{\alpha_1} - 1)/\alpha_1]$.

3. $m = 3$ with $c_{ijk} = \varepsilon_{ijk}$, the Levi–Civita symbol, the Lie groups $SU(2)$ and $SO(3)$. That such an array of structure constants satisfies the quadratic Lie identities (B.10) is verified immediately by recalling the relation $\sum_{h=1}^{3} \varepsilon_{ijh} \varepsilon_{hkl} = \delta_{ik} \delta_{jl} - \delta_{il} \delta_{jk}$. With $c_{ijk} = \varepsilon_{ijk}$, generator equations (B.8) become

$$G_1 G_2 - G_2 G_1 = G_3$$
$$G_2 G_3 - G_3 G_2 = G_1 \tag{B.15}$$
$$G_3 G_1 - G_1 G_3 = G_2.$$

We have the two-dimensional matrix representation

$$G_1 = \frac{1}{2} \begin{pmatrix} 0 & -i \\ -i & 0 \end{pmatrix}, \qquad G_2 = \frac{1}{2} \begin{pmatrix} 0 & -1 \\ 1 & 0 \end{pmatrix}, \qquad G_3 = \frac{1}{2} \begin{pmatrix} -i & 0 \\ 0 & i \end{pmatrix},$$

$$\tag{B.16}$$

as well as the three-dimensional adjoint representation

$$G_1 = \begin{pmatrix} 0 & 0 & 0 \\ 0 & 0 & -1 \\ 0 & 1 & 0 \end{pmatrix}, \qquad G_2 = \begin{pmatrix} 0 & 0 & 1 \\ 0 & 0 & 0 \\ -1 & 0 & 0 \end{pmatrix},$$

$$G_3 = \begin{pmatrix} 0 & -1 & 0 \\ 1 & 0 & 0 \\ 0 & 0 & 0 \end{pmatrix}. \tag{B.17}$$

By putting the two-dimensional representation (B.16) into (B.11), we obtain

$$T(\alpha) = \exp \frac{1}{2} \begin{pmatrix} -i\alpha_3 & -i\alpha_1 - \alpha_2 \\ -i\alpha_1 + \alpha_2 & i\alpha_3 \end{pmatrix}$$

$$= \left(\cos \frac{1}{2}|\alpha|\right)\begin{pmatrix} 1 & 0 \\ 0 & 1 \end{pmatrix} - \frac{i(\sin \frac{1}{2}|\alpha|)}{|\alpha|}\begin{pmatrix} \alpha_3 & \alpha_1 - i\alpha_2 \\ \alpha_1 + i\alpha_2 & -\alpha_3 \end{pmatrix}, \tag{B.18}$$

$$|\alpha| \equiv (\alpha_1{}^2 + \alpha_2{}^2 + \alpha_3{}^2)^{1/2},$$

a parametric representation of all 2×2 unitary matrices of determinant one, the Lie group $SU(2)$.[4] All elements in the group are included if α is restricted to a sphere about the origin of radius 2π, $|\alpha| \leqslant 2\pi$, with all points on the spherical surface $|\alpha| = 2\pi$ being identified with the group element

$$\begin{pmatrix} -1 & 0 \\ 0 & -1 \end{pmatrix};$$

thus, $SU(2)$ is compact and simply connected. By putting the three-dimensional adjoint representation (B.17) into (B.11), we obtain

$$T(\alpha) = \exp\begin{pmatrix} 0 & -\alpha_3 & \alpha_2 \\ \alpha_3 & 0 & -\alpha_1 \\ -\alpha_2 & \alpha_1 & 0 \end{pmatrix}, \tag{B.19}$$

a parametric representation of all 3×3 real orthogonal matrices of determinant one, the Lie group $SO(3)$.[4] All elements in the group are

---

[4] The nomenclature symbol S stands for "special" and means "determinant equal to one." The determinant of an exponentiated traceless matrix [such as (B.18), (B.19), or (B.26)] equals one because we have

$$\det(\exp \mathbf{M}) = \lim_{N \to \infty} \det((1 + N^{-1}\mathbf{M})^N)$$
$$= \lim_{N \to \infty} (\det(1 + N^{-1}\mathbf{M}))^N$$
$$= \lim_{N \to \infty} (1 + N^{-1}(\operatorname{tr}\mathbf{M}) + O(N^{-2}))^N$$
$$= \exp(\operatorname{tr}\mathbf{M})$$

included if α is restricted to a sphere about the origin of radius π, $|\alpha| \equiv (\alpha_1{}^2 + \alpha_2{}^2 + \alpha_3{}^2)^{1/2} \leqslant \pi$, with each pair of opposite points on the spherical surface $|\alpha| = \pi$ being identified with a single group element (which squares to the identity because $T(\alpha)T(-\alpha) = \mathbf{1}$); hence, $SO(3)$ is compact and doubly connected. Since linear representations of both $SU(2)$ and $SO(3)$ follow from the structure constants $c_{ijk} = \varepsilon_{ijk}$, the Lie algebras for $SU(2)$ and $SO(3)$ are isomorphic, which implies that the Lie groups $SU(2)$ and $SO(3)$ are isomorphic in the neighborhood of the identity. On the other hand, the appropriate parameter set domains and topological character (manifest by the identification of points with group elements on the critical spherical surfaces) are different for the complete sets of group elements, and so the isomorphism between the Lie groups $SU(2)$ and $SO(3)$ is local but not global. We also note that the differential operator solution to (B.15),

$$G_1 = -x_2 \frac{\partial}{\partial x_3} + x_3 \frac{\partial}{\partial x_2}, \qquad G_2 = -x_3 \frac{\partial}{\partial x_1} + x_1 \frac{\partial}{\partial x_3},$$

$$G_3 = -x_1 \frac{\partial}{\partial x_2} + x_2 \frac{\partial}{\partial x_1}, \tag{B.20}$$

the associated adjoint representation, yields the set of differential operator group elements

$$T(\alpha) = \exp\left(-\sum_{i,\,j,\,k=1}^{3} \varepsilon_{ijk}\alpha_i x_j (\partial/\partial x_k)\right) \tag{B.21}$$

that rotates the 3-tuple argument of a C^∞ 1-tuple function $f(x)$, $T(\alpha)f(x) = f(x')$, where x' is related to x by the inverse of the associated $SO(3)$ matrix of the form (B.19),

$$\begin{pmatrix} x_1' \\ x_2' \\ x_3' \end{pmatrix} = \left[\exp\begin{pmatrix} 0 & \alpha_3 & -\alpha_2 \\ -\alpha_3 & 0 & \alpha_1 \\ \alpha_2 & -\alpha_1 & 0 \end{pmatrix} \right] \begin{pmatrix} x_1 \\ x_2 \\ x_3 \end{pmatrix}. \tag{B.22}$$

4. $m = 3$ with $c_{ijk} = \varepsilon_{ijk}\tau_k$, $\tau_k \equiv (-1)^{k-1}$, the Lie group $SL(2, R)$. To show that this array of structure constants satisfies the quadratic Lie identities (B.10), we first establish the formula

$$\sum_{h=1}^{3} c_{ijh} c_{hkl} = (-1)^k(\delta_{ik}\,\delta_{jl} - \delta_{il}\,\delta_{jk}), \tag{B.23}$$

from which (B.10) follows by cyclic permutation of i, j, k and addition. The generator equations (B.8),

$$G_1 G_2 - G_2 G_1 = G_3$$
$$G_2 G_3 - G_3 G_2 = G_1 \qquad (\text{B.24})$$
$$G_3 G_1 - G_1 G_3 = -G_2,$$

admit the two-dimensional matrix representation

$$G_1 = \frac{1}{2}\begin{pmatrix} 0 & 1 \\ 1 & 0 \end{pmatrix}, \qquad G_2 = \frac{1}{2}\begin{pmatrix} 0 & -1 \\ 1 & 0 \end{pmatrix}, \qquad G_3 = \frac{1}{2}\begin{pmatrix} 1 & 0 \\ 0 & -1 \end{pmatrix}. \qquad (\text{B.25})$$

By putting the latter set of generators into (B.11), we obtain

$$T(\alpha) = \exp \frac{1}{2}\begin{pmatrix} \alpha_3 & \alpha_1 - \alpha_2 \\ \alpha_1 + \alpha_2 & -\alpha_3 \end{pmatrix}$$

$$= \left(\cosh \frac{1}{2}(\alpha_1{}^2 - \alpha_2{}^2 + \alpha_3{}^2)^{1/2} \right)\begin{pmatrix} 1 & 0 \\ 0 & 1 \end{pmatrix}$$

$$+ \left(\frac{\sinh \frac{1}{2}(\alpha_1{}^2 - \alpha_2{}^2 + \alpha_3{}^2)^{1/2}}{(\alpha_1{}^2 - \alpha_2{}^2 + \alpha_3{}^2)^{1/2}} \right)\begin{pmatrix} \alpha_3 & \alpha_1 - \alpha_2 \\ \alpha_1 + \alpha_2 & -\alpha_3 \end{pmatrix}$$

$$\text{for } (\alpha_1{}^2 - \alpha_2{}^2 + \alpha_3{}^2) \geqslant 0,$$

$$= \left(\cos \frac{1}{2}(-\alpha_1{}^2 + \alpha_2{}^2 - \alpha_3{}^2)^{1/2} \right)\begin{pmatrix} 1 & 0 \\ 0 & 1 \end{pmatrix}$$

$$+ \left(\frac{\sin \frac{1}{2}(-\alpha_1{}^2 + \alpha_2{}^2 - \alpha_3{}^2)^{1/2}}{(-\alpha_1{}^2 + \alpha_2{}^2 - \alpha_3{}^2)^{1/2}} \right)\begin{pmatrix} \alpha_3 & \alpha_1 - \alpha_2 \\ \alpha_1 + \alpha_2 & -\alpha_3 \end{pmatrix}$$

$$\text{for } (\alpha_1{}^2 - \alpha_2{}^2 + \alpha_3{}^2) \leqslant 0,$$

$$(\text{B.26})$$

a parametric representation of 2×2 real matrices of determinant one, a part of the Lie group $SL(2, R)$.[5] All elements of $SL(2, R)$ are included in the augmented set $\{(\exp \alpha \cdot G), -(\exp \alpha \cdot G)\}$ with the components of α assuming all finite real values for which $(\alpha_1{}^2 - \alpha_2{}^2 + \alpha_3{}^2) \geqslant -\pi^2$ with opposite points on the hyperboloid surfaces

$$\alpha_2 = \pm(\alpha_1{}^2 + \alpha_3{}^2 + \pi^2)^{1/2}$$

[5] The trace of matrices of the form (B.26) is greater or equal to -2; to get the matrices with trace less than -2, we must also consider the derived set $\{-(\exp \alpha \cdot G)\}$.

identified with a traceless matrix that occurs in both subsets $\{(\exp \alpha \cdot G)\}$ and $\{-(\exp \alpha \cdot G)\}$; thus, $SL(2, R)$ is noncompact and has the connectivity of a periodic slab. By making the formal parameter replacement $(\alpha_1, \alpha_2, \alpha_3) \rightarrow (-i\alpha_1, \alpha_2, -i\alpha_3)$, one obtains (B.18) from (B.26), and so $SU(2)$ is the (unique) *compact complex-extension* of $SL(2, R)$.

An immediate general classification of m-parameter Lie groups follows from *Cartan's metric*, defined in terms of the structure constants as the symmetric $m \times m$ array

$$g_{ij} \equiv \sum_{k,l=1}^{m} c_{ikl} c_{jlk}. \tag{B.27}$$

We have $(g_{ij}) = 0$ for the m-dimensional Abelian Lie group (Example 1 above), $(g_{ij}) = \mathrm{diag}\ulcorner 1, 0 \lrcorner$ for the proper affine group on real numbers (Example 2 above), $(g_{ij}) = \mathrm{diag}\ulcorner -2, -2, -2 \lrcorner$ for the Lie groups $SU(2)$ and $SO(3)$ (Example 3 above), and $(g_{ij}) = \mathrm{diag}\ulcorner 2, -2, 2 \lrcorner$ for the Lie group $SL(2, R)$ (Example 4 above). Cartan's metric (B.27) is negative-definite for compact Lie groups and indefinite for noncompact Lie groups. By making use of (B.10), we find that the quantity

$$a_{ijk} \equiv \sum_{h=1}^{m} c_{ijh} g_{hk} = \sum_{r,s,t=1}^{m} (c_{ris} c_{sjt} c_{tkr} - c_{rjs} c_{sit} c_{tkr})$$
$$\equiv -a_{ikj} \equiv a_{kij} \equiv -a_{kji} \tag{B.28}$$

is, in general, a totally antisymmetric array of real constants, changing its sign under transposition of any two indices. An m-parameter Lie group is *semisimple* if the $m \times m$ array (B.27) constitutes a nonsingular matrix,

$$\det(g_{ij}) \neq 0. \tag{B.29}$$

Thus, the m-dimensional Abelian Lie group and the proper affine Lie group on real numbers are not semisimple, while the Lie groups $SU(2)$, $SO(3)$, and $SL(2, R)$ are semisimple. The symmetric matrix inverse of Cartan's metric (B.27), customarily denoted g^{ij}, exists for a semisimple Lie group and satisfies

$$\sum_{k=1}^{m} g^{ik} g_{kj} = \delta_j^{\ i}. \tag{B.30}$$

In terms of g^{ij} and the generators for a linear representation of a semisimple Lie group, the *quadratic Casimir invariant* is defined as

$$C \equiv \sum_{i,j=1}^{m} g^{ij} G_i G_j. \tag{B.31}$$

Because it commutes with all the generators,

$$CG_k - G_k C = \sum_{i,j,h=1}^{m} g^{ij}(c_{jkh} G_i G_h + c_{ikh} G_h G_j)$$

$$= \sum_{i,j,h=1}^{m} (g^{ij} c_{ikh} + g^{ih} c_{ikj}) G_j G_h$$

$$= \sum_{i,j,h,l=1}^{m} (g^{ij} g^{hl} + g^{ih} g^{jl}) a_{ikl} G_j G_h$$

$$= \sum_{i,j,h,l=1}^{m} g^{ij} g^{hl}(a_{ikl} + a_{lki}) G_j G_h = 0 \qquad (B.32)$$

as a consequence of the antisymmetry of (B.28), the quadratic Casimir invariant plays a central role in the representation theory for semi-simple Lie groups.[6]

---

[6] For general discussions of Lie group representation theory, see L. S. Pontrjagin, "Topological Groups." Princeton Univ. Press, Princeton, New Jersey, 1946; R. E. Behrends, *et al.*, *Rev. Modern Phys.* **34**, 1 (1962) and works cited therein.

Appendix C *Haar Measure*

Functions of the linear representations of Lie groups discussed in Appendix B can be "averaged over the group" by defining appropriate integrals with respect to the m-tuple parameter α. There is a general method for the computation of an infinitesimal volume element in $\mathscr{G}$, denoted by $D(T(\alpha))$ and equal to (an appropriate nonnegative 1-tuple function of α) $\times \prod_{k=1}^{m} d\alpha_k$, for which integrals over $\mathscr{G}$ are left-invariant,

$$\int_{\mathscr{G}} \Phi(T(\alpha))D_L(T(\alpha)) \equiv \int_{\mathscr{G}} \Phi(T(\beta)T(\alpha))D_L(T(\alpha)) \tag{C.1}$$

for all $T(\beta) \in \mathscr{G}$, or right-invariant,

$$\int_{\mathscr{G}} \Phi(T(\alpha))D_R(T(\alpha)) \equiv \int_{\mathscr{G}} \Phi(T(\alpha)T(\beta))D_R(T(\alpha)) \tag{C.2}$$

for all $T(\beta) \in \mathscr{G}$. The quantities $\Phi(T(\alpha))$ in (C.1) and (C.2), representing generic functions of the group elements, can be numerical-valued, n-tuple-valued, or operator-valued, the *Hurwitz integrals*[1] (C.1) and (C.2) being defined in the range of $\Phi(T(\alpha))$. From the fact that $\mathscr{G}$ is a

---

[1] For a penetrating discussion of the Hurwitz integrals, see E. P. Wigner, "Group Theory," pp. 95–101. Academic Press, New York, 1959.

group, the left-invariant integral (C.1) requires an infinitesimal volume element such that

$$D_L(T(\alpha)) \equiv D_L(T(\beta)T(\alpha)) \qquad (C.3)$$

for all fixed $T(\beta) \in \mathscr{G}$, while the right-invariant integral (C.2) requires an infinitesimal volume element such that

$$D_R(T(\alpha)) \equiv D_R(T(\alpha)T(\beta)) \qquad (C.4)$$

for all fixed $T(\beta) \in \mathscr{G}$. An infinitesimal volume element with the property (C.3) or (C.4) is a left-invariant or right-invariant *Haar measure*[2] on $\mathscr{G}$.

To obtain general formulas for the left-invariant Haar measure associated with a linear representation of an m-parameter Lie group, first differentiate (B.4) with respect to β_i, set $\beta = 0$, and make use of Eqs. (B.2) and (B.5); the m operator equations that result are

$$T(\alpha)G_i = \sum_{j=1}^{m} \rho_{ij}(\alpha)[\partial T(\alpha)/\partial \alpha_j] \qquad (C.5)$$

with the $m \times m$ array $\rho_{ij}(\alpha) \equiv \partial \gamma_j/\partial \beta_i|_{\beta=0}$. Next, multiply (C.5) by $T(\bar{\gamma})$ from the left and make use of the relation derived from (B.3) and (B.4), $T(\bar{\gamma})T(\alpha) = T(\bar{\beta})$; the resulting equations are

$$T(\bar{\beta})G_i = \sum_{j,k=1}^{m} \rho_{ij}(\alpha)(\partial\bar{\beta}_k/\partial\alpha_j)[\partial T(\bar{\beta})/\partial\bar{\beta}_k] \qquad (C.6)$$

where α and $\bar{\gamma}$ are viewed as independent m-tuples with $\bar{\beta} = \bar{\beta}(\alpha, \bar{\gamma})$ prescribed by the group multiplication. Finally, replace α by $\bar{\beta}$ in (C.5) and compare these equations with (C.6) to obtain

$$\sum_{j=1}^{m} \rho_{ij}(\alpha)(\partial\bar{\beta}_k/\partial\alpha_j) = \rho_{ik}(\bar{\beta}) \qquad (C.7)$$

for all $\bar{\gamma}$ independent of α. In view of (C.7), the left-invariant Haar measure is defined to within normalization by

$$D_L(T(\alpha)) \equiv [\det(\rho_{ij}(\alpha))]^{-1} \prod_{k=1}^{m} d\alpha_k, \qquad (C.8)$$

---

[2] The classic abstract discussion of Haar measure is that of P. R. Halmos "Measure Theory," pp. 250–265. Van Nostrand, Princeton, New Jersey, 1950.

for it follows from (C.8) and (C.7) that

$$D_L(T(\bar{\gamma})T(\alpha)) = D_L(T(\bar{\beta})) \equiv [\det(\rho_{ij}(\bar{\beta}))]^{-1} \prod_{k=1}^{m} d\bar{\beta}_k$$

$$= [\det(\rho_{ij}(\alpha))]^{-1} [\det(\partial\bar{\beta}_i/\partial\alpha_j)]^{-1} \prod_{k=1}^{m} d\bar{\beta}_k$$

$$= D_L(T(\alpha)) \tag{C.9}$$

for all $\bar{\gamma}$ independent of α. A similar calculation produces the formula for the right-invariant Haar measure,

$$D_R(T(\alpha)) \equiv [\det(\omega_{ij}(\alpha))]^{-1} \prod_{k=1}^{m} d\alpha_k, \tag{C.10}$$

to within normalization and with the $m \times m$ array $\omega_{ij}(\alpha)$ given implicitly by

$$G_i T(\alpha) = \sum_{j=1}^{m} \omega_{ij}(\alpha)[\partial T(\alpha)/\partial\alpha_j]. \tag{C.11}$$

The left-invariant and right-invariant Haar measures (C.8) and (C.10) are related by the equation $D_L(T(\alpha)) = D_R(T(\bar{\alpha}))$ where $T(\bar{\alpha}) = T(\alpha)^{-1}$ since we have

$$\omega_{ij}(\bar{\alpha}) = -\sum_{k=1}^{m} \rho_{ik}(\alpha)\partial\bar{\alpha}_j/\partial\alpha_k, \tag{C.12}$$

because of (C.5) and (C.11). If $\mathcal{G}$ is compact, the left-invariant and right-invariant Haar measures (C.8) and (C.10) are identical.[2]

The Haar measure formulas in the preceding paragraph do not depend on the parametric labeling of the elements of $\mathcal{G}$. Subject to a parameter transformation $\alpha \to \alpha' = \alpha +$ (terms quadratic and of higher order in α) which relabels the group elements with $T'(\alpha') \equiv T(\alpha)$, both Haar measures (C.8) and (C.10) transform as scalar invariants, $D'(T'(\alpha')) = D(T(\alpha))$, as a consequence of the $\rho_{ij}(\alpha)$ and $\omega_{ij}(\alpha)$ parameter transformation character implied by (C.5) and (C.11). However, the actual computation of $\rho_{ij}(\alpha)$ and $\omega_{ij}(\alpha)$ is simplified for $\mathcal{G}$ with $m \geqslant 3$ by employing the parametric labeling for which the elements of $\mathcal{G}$ are given by the canonical expression (B.11). Then the formula for the partial derivatives of an exponential operator function[3] applies and produces[4]

$$(\partial/\partial\alpha_i)(\exp \alpha \cdot G) = \int_0^1 (\exp(1-s)\alpha \cdot G)G_i(\exp s\alpha \cdot G)\, ds$$

$$\equiv \int_0^1 (\exp s\alpha \cdot G)G_i(\exp(1-s)\alpha \cdot G)\, ds, \quad (C.13)$$

where s is a dummy 1-tuple integration variable. Hence, by putting (B.11) into (C.5) and (C.11), we have the more explicit equations for $\rho_{ij}(\alpha)$ and $\omega_{ij}(\alpha)$,

$$\sum_{j=1}^m \rho_{ij}(\alpha) \int_0^1 (\exp -s\alpha \cdot G)G_j(\exp s\alpha \cdot G)\, ds$$

$$= G_i = \sum_{j=1}^m \omega_{ij}(\alpha) \int_0^1 (\exp s\alpha \cdot G)G_j(\exp -s\alpha \cdot G)\, ds. \quad (C.14)$$

Haar measures for the linear representations of Lie groups mentioned in Appendix B are obtained to within normalization as follows:

1. For the m-dimensional Abelian Lie group with all G_i commuting and the elements of $\mathscr{G}$ expressed canonically as (B.12), we have

$$\partial T(\alpha)/\partial\alpha_i = G_i T(\alpha) = T(\alpha)G_i. \quad (C.15)$$

Hence, both $m \times m$ arrays $\rho_{ij}(\alpha)$ in (C.5) and $\omega_{ij}(\alpha)$ in (C.11) equal δ_{ij}, and so the left-invariant and right-invariant Haar measures (C.8) and (C.10) are given by

$$D(T(\alpha)) = \prod_{k=1}^m d\alpha_k. \quad (C.16)$$

2. For the proper affine Lie group on real numbers with the representation (B.14), we find

[3] For example, R. M. Wilcox, *J. Math. Phys.* **8**, 962 (1967) and works cited therein.

[4] A direct proof of (C.13) follows from the equation

$$\left(\frac{\partial}{\partial\lambda} - (\alpha \cdot G)\right)\left[\frac{\partial}{\partial\alpha_i}\left(\exp \lambda\alpha \cdot G\right) - \int_0^1 \left(\exp(1-s)\lambda\alpha \cdot G\right)\lambda G_i\left(\exp s\lambda\alpha \cdot G\right)ds\right] = 0$$

where λ is a real 1-tuple parameter; since the square-bracketed quantity in the latter equation vanishes for $\lambda = 0$, the equation implies that the square-bracketed quantity vanishes for all λ and, in particular, for $\lambda = 1$.

$$(\rho_{ij}(\alpha)) = \begin{pmatrix} 1 & \alpha_2[\alpha_1^{-1} - (1 - e^{-\alpha_1})^{-1}] \\ 0 & \alpha_1(1 - e^{-\alpha_1})^{-1} \end{pmatrix}$$

by working out terms in (C.5) or the left member of (C.14). Hence, the left-invariant Haar measure (C.8) is

$$D_L(T(\alpha)) = [(1 - e^{-\alpha_1})/\alpha_1] \, d\alpha_1 \, d\alpha_2. \tag{C.17}$$

On the other hand, by working out terms in (C.11) or the right member (C.14), we find

$$(\omega_{ij}(\alpha)) = \begin{pmatrix} 1 & \alpha_2[\alpha_1^{-1} - (e^{\alpha_1} - 1)^{-1}] \\ 0 & \alpha_1(e^{\alpha_1} - 1)^{-1} \end{pmatrix}.$$

Hence, the right-invariant Haar measure (C.10) is

$$D_R(T(\alpha)) = [(e^{\alpha_1} - 1)/\alpha_1] \, d\alpha_1 \, d\alpha_2. \tag{C.18}$$

Note that the Haar measures (C.17) and (C.18) are different for this $\mathscr{G}$.

3. For the Lie groups $SU(2)$ and $SO(3)$ with the representations (B.18) and (B.19), the calculations are facilitated by making use of the fact that α_1, α_2, α_3 enter with rotational symmetry.[5] Since the weight factors $[\det(\rho_{ij}(\alpha))]^{-1}$ and $[\det(\omega_{ij}(\alpha))]^{-1}$ in (C.8) and (C.10) can depend only on $|\alpha| \equiv (\alpha_1^2 + \alpha_2^2 + \alpha_3^2)^{1/2}$, we set $\alpha = (0, 0, |\alpha|)$ in (C.14) and

---

[5] It is not difficult to demonstrate the invariance of (C.8) with respect to proper rotations of the 3-tuple parameter $\alpha = (\alpha_1, \alpha_2, \alpha_3)$ in the $SU(2)$ representation (B.18). To show that $D_L(T(\alpha')) = D_L(T(\alpha))$ for $\alpha_i' = R_{ij}\alpha_j$ with $R_{ij}R_{ik} = \delta_{jk}$, $\det(R_{ij}) = +1$, we first associate the proper rotation matrix (R_{ij}) (an element of $SO(3)$) with a 2×2 unitary matrix of determinant one $T(\beta)$ (a fixed element of $SU(2)$) by solving the equations

$$T(\beta)G_i = R_{ij}G_jT(\beta).$$

where the generators are given by (B.16) as the Pauli matrices times the numerical factor $(-i/2)$; the latter equations prescribe $T(\beta)$ in terms of (R_{ij}) uniquely to within a (± 1) numerical factor and imply that

$$T(\beta)T(\alpha') = T(\beta)(\exp R_{ij}\alpha_j G_i) = T(\alpha)T(\beta).$$

Next, we replace α by α' in (C.5) and move $T(\beta)$ through the equation from left to right to obtain

$$R_{ij}T(\alpha)G_j = \rho_{ij}(\alpha')[\partial T(\alpha)/\partial\alpha_j']$$

from which it follows that

$$\rho_{ij}(\alpha') = R_{ik}R_{jl}\rho_{kl}(\alpha),$$

solve for the weighting factors associated with the $SU(2)$ representation (B.18),

$$[\det(\rho_{ij}(\alpha))]^{-1} = [\det(m_{ij}(\alpha))]^{-1} = (2(\sin \tfrac{1}{2}|\alpha|)/|\alpha|)^2. \qquad (C.19)$$

As a consequence of the local isomorphism of $SU(2)$ and $SO(3)$, the identical weighting factor is obtained by solving (C.14) with the $SO(3)$ representations (B.19) and (B.21). Thus the left-invariant and right-invariant Haar measure, the same for the linear representations of $SU(2)$ and $SO(3)$, is given by

$$D(T(\alpha)) = (2(\sin \tfrac{1}{2}|\alpha|)/|\alpha|)^2 \, d\alpha_1 \, d\alpha_2 \, d\alpha_3. \qquad (C.20)$$

Observe that the Haar measure (C.20) vanishes on the critical spherical surface $|\alpha| = 2\pi$ for $SU(2)$, but is finite on the critical spherical surface $|\alpha| = \pi$ for $SO(3)$.

4. For the Lie group $SL(2, R)$ with the representation (B.26), the calculations can be related directly to the $SU(2)$ calculations in the preceding example by making the analytic continuation which takes (B.18) into (B.26), namely: $\alpha_1 \to i\alpha_1$, $\alpha_2 \to \alpha_2$, $\alpha_3 \to i\alpha_3$. By applying this analytic continuation to (C.20), we find that the invariant Haar measures are equal and given by

$$D(T(\alpha)) \equiv D(\pm(\exp \alpha \cdot G))$$

$$= \left(\frac{2 \sin \tfrac{1}{2}(-\alpha_1^{\,2} + \alpha_2^{\,2} - \alpha_3^{\,2})^{1/2}}{(-\alpha_1^{\,2} + \alpha_2^{\,2} - \alpha_3^{\,2})^{1/2}} \right)^2 d\alpha_1 \, d\alpha_2 \, d\alpha_3$$

$$\text{for} \qquad 0 \leqslant (-\alpha_1^{\,2} + \alpha_2^{\,2} - \alpha_3^{\,2}) \leqslant \pi^2$$

$$= \left(\frac{2 \sinh \tfrac{1}{2}(\alpha_1^{\,2} - \alpha_2^{\,2} + \alpha_3^{\,2})^{1/2}}{(\alpha_1^{\,2} - \alpha_2^{\,2} + \alpha_3^{\,2})^{1/2}} \right)^2 d\alpha_1 \, d\alpha_2 \, d\alpha_3$$

$$\text{for} \qquad (\alpha_1^{\,2} - \alpha_2^{\,2} + \alpha_3^{\,2}) \geqslant 0. \qquad (C.21)$$

and hence that $\det(\rho_{ij}(\alpha')) = \det(\rho_{ij}(\alpha))$. Thus, we have

$$D_L(T(\alpha')) \equiv [\det(\rho_{ij}(\alpha'))]^{-1} \prod_{k=1}^{3} d\alpha_k' = D_L(T(\alpha)).$$

It is interesting to note that explicit computation with (C.5) can be avoided if we use the fact that $D_L(T(\alpha)) = D_R(T(\alpha))$, because of $SU(2)$ being compact[2]; we have $D_L(T(\alpha')) = D_L(T(\beta)T(\alpha')) = D_L(T(\alpha)T(\beta)) = D_R(T(\alpha)T(\beta)) = D_R(T(\alpha)) \equiv D_L(T(\alpha))$.

Appendix D Functional Integration by Parts Lemma

Here we establish the lemma[1]: *If the functional derivative $\delta F/\delta f(x)$ and the functional integral $\int_{\mathscr{F}} F D(f)$ both exist for a certain functional $F = F[f(x)]$ and function space $\mathscr{F}$ with the measure displacement-invariant $(D(f + \sigma) = D(f)$ for any fixed $\sigma \in \mathscr{F})$ then*

$$\int_{\mathscr{F}} [\delta F/\delta f(x)]D(f) = 0. \qquad (D.1)$$

The notation is that of Appendix A. To prove the lemma, we observe that

$$\int_{\mathscr{F}} F[f(x) + \sigma(x)]D(f) = \int_{\mathscr{F}} F[f(x) + \sigma(x)]D(f + \sigma) = \int_{\mathscr{F}} F[f(x)]D(f)$$
$$(D.2)$$

with the first equality in (D.2) a consequence of the measure being displacement-invariant, and the second equality in (D.2) the result of replacing the dummy integration variable, $(f + \sigma) \to f$. Since the last member of (D.2) is independent of σ, (D.1) follows by taking the functional derivative of (D.2) with respect to $\sigma(x)$ and setting $\sigma = 0$.

[1] G. Rosen, *Phys. Rev. Lett.* **16**, 704 (1966), and works cited therein.

To illustrate the utility of the lemma, let us use it to derive the Wiener correlation function[2]

$$b(t', t''; t) \equiv \int_{\mathscr{F}} f(t')f(t'') \, \Delta(f) = \tfrac{1}{2} \min(t', t'') \qquad \text{for} \quad 0 \leqslant t', t'' \leqslant t,$$

(D.3)

where $\mathscr{F} = \{f = f(\tau) \text{ for } 0 \leqslant \tau \leqslant t: f(\tau) \in C^0, f(0) = 0\}$ with τ a real 1-tuple variable; the *Wiener measure* in (D.3),

$$\Delta(f) \equiv \left(\exp - \int_0^t (df(\tau)/d\tau)^2 \, d\tau \right) D(f) \qquad \text{(D.4)}$$

with $D(f + \sigma) \equiv D(f)$ displacement-invariant for any fixed $\sigma \in \mathscr{F}$ is normalized to produce

$$\int_{\mathscr{F}} \Delta(f) \equiv 1. \qquad \text{(D.5)}$$

We have

$$\int_{\mathscr{F}} f(t') \, \Delta(f) = 0 \qquad \text{for} \quad 0 \leqslant t' \leqslant t, \qquad \text{(D.6)}$$

because the integrand in (D.6), $F = F[f(\tau)] = f(t') \exp - \int_0^t (df(\tau)/d\tau)^2 \, d\tau$, is an odd functional of $f(\tau)$. By applying the functional integration by parts lemma to this F, it follows that $\int_{\mathscr{F}} \delta F/\delta f(t'') D(f) = 0$, and we obtain

$$\int_{\mathscr{F}} (\delta(t' - t'') + 2f(t')[d^2 f(t'')/dt''^2]) \, \Delta(f) = 0 \qquad \text{for} \quad 0 \leqslant t', t'' \leqslant t.$$

(D.7)

In view of (D.5), (D.7) becomes

$$\int_{\mathscr{F}} f(t')(d^2 f(t'')/dt''^2) \, \Delta(f) = -\tfrac{1}{2} \delta(t' - t'') \qquad \text{for} \quad 0 \leqslant t', t'' \leqslant t,$$

(D.8)

which implies that

$$\int_{\mathscr{F}} f(t')(df(s)/ds) \, \Delta(f) = \begin{array}{ll} \tfrac{1}{2} & \text{for} \quad 0 \leqslant s < t' \leqslant t \\ 0 & \text{for} \quad 0 \leqslant t' \leqslant s \leqslant t, \end{array} \qquad \text{(D.9)}$$

because the left-hand side of (D.9) vanishes if $t' = 0$. The expression (D.3) for the Wiener correlation function follows by integrating (D.9) with respect to s from 0 to t''

[2] For example, see I. M. Gel'fand and A. M. Yaglom, *J. Math. Phys.* **1**, 48 (1960).

Relativistic Sum-over-Histories
for the One-Dimensional
Dirac Equation

Heuristic interest is attached to the one-dimensional Dirac equation[1]

$$(i(\partial/\partial t) + i\alpha(\partial/\partial x) - \beta)\psi(x; t) = 0, \tag{E.1}$$

where $\psi(x; t)$ is a two-component wave function, x is a real 1-tuple coordinate, and the 2×2 Hermitian matrices α and β satisfy the conditions

$$\alpha^2 = \beta^2 = \begin{pmatrix} 1 & 0 \\ 0 & 1 \end{pmatrix}, \qquad \alpha\beta + \beta\alpha = \begin{pmatrix} 0 & 0 \\ 0 & 0 \end{pmatrix}. \tag{E.2}$$

For a prescribed initial state $\psi(x; 0) \equiv \psi_0(x)$, the solution to (E.1) involves the propagation kernel

$$\psi(x; t) = \int_{-\infty}^{\infty} K(x - x', t)\psi_0(x') \, dx' \tag{E.3}$$

involves the propagation kernel

$$K(x, t) = \left(\frac{\partial}{\partial t} - \alpha\frac{\partial}{\partial x} - i\beta\right)[J_0((t^2 - x^2)^{1/2})\sigma(t - |x|)]$$

$$\sigma(u) \equiv \begin{matrix} 1 & u > 0 \\ \tfrac{1}{2} & u = 0 \\ 0 & u < 0. \end{matrix} \tag{E.4}$$

---

[1] Physical units are chosen such that the speed of light, the mass of the electron, and $\hbar$ equal one.

That (E.4) satisfies the equations derived from (E.1) and (E.3),

$$(i(\partial/\partial t) + i\alpha(\partial/\partial x) - \beta)K(x, t) = 0 \tag{E.5}$$

$$\lim_{t \to 0} K(x, t) = \begin{pmatrix} \delta(x) & 0 \\ 0 & \delta(x) \end{pmatrix}, \tag{E.6}$$

is verified immediately by evoking the zero-order Bessel function representation[2]

$$J_0((t^2 - x^2)^{1/2})\sigma(t - |x|) = \frac{1}{\pi} \int_{-\infty}^{\infty} \frac{[\sin(k^2 + 1)^{1/2}t][\cos kx]}{(k^2 + 1)^{1/2}} \, dk \tag{E.7}$$

for $t > 0$. On the other hand, by substituting the Maclaurin series representation for J_0 into (E.4), we find that

$$K(x, t) = \sum_{n=0}^{\infty} (-1)^{n+1} \left(\frac{(t + \alpha x)}{2^{2n+1}(n+1)!\,n!} + \frac{i\beta}{2^{2n}(n!)^2} \right)(t^2 - x^2)^n \tag{E.8}$$

for $t > |x|$.

Feynman[3] has noted a remarkable "sum-over-histories" representation for this propagation kernel, namely:

$$K(x, t) = \lim_{\varepsilon \to 0} \mathscr{K}(x/\varepsilon, t/\varepsilon; \varepsilon), \tag{E.9}$$

$$\mathscr{K}(r, s; \varepsilon) \equiv \varepsilon^{-1} \sum_{R=0}^{\infty} \begin{pmatrix} N_{++}(r, s; R) & N_{+-}(r, s; R) \\ N_{-+}(r, s; R) & N_{--}(r, s; R) \end{pmatrix} (i\varepsilon)^R, \tag{E.10}$$

in which $N_{\xi\eta}(r, s; R)$ denotes the number of "histories" composed of s unit steps from the origin to the (positive or negative) integer r [$\frac{1}{2}(s + r)$ unit steps forward with $\Delta r = +1$ and $\frac{1}{2}(s - r)$ unit steps backward with $\Delta r = -1$] that have R "reversals" (that is, changes in the sign of successive Δr) and where η is the sign of the initial step and ξ is the sign of the final step. Thus, at any instant of time the electron

[2] I. S. Gradshteyn and I. M. Ryzhik, "Table of Integrals, Series, and Products," p. 472. Academic Press, New York, 1965.

[3] R. P. Feynman and A. R. Hibbs, "Quantum Mechanics and Path Integrals," p. 35. McGraw-Hill, New York, 1965.

is considered to move forward or backward at the speed of light, with the spin state associated with the instantaneous velocity.[4] From the definition of the N's, it follows that[5]

$$N_{++}(r, s + 1; R) = N_{++}(r - 1, s; R) + N_{-+}(r - 1, s; R - 1)$$

$$N_{+-}(r, s + 1; R) = N_{+-}(r - 1, s; R) + N_{--}(r - 1, s; R - 1)$$

$$N_{-+}(r, s + 1; R) = N_{++}(r + 1, s; R - 1) + N_{-+}(r + 1, s; R)$$ (E.11)

$$N_{--}(r, s + 1; R) = N_{+-}(r + 1, s; R - 1) + N_{--}(r + 1, s; R)$$

for all $s \geqslant 1$, and hence the definition (E.10) implies that

$$\mathcal{K}(r, s + 1; \varepsilon) = \begin{pmatrix} 1 & i\varepsilon \\ 0 & 0 \end{pmatrix} \mathcal{K}(r - 1, s; \varepsilon) + \begin{pmatrix} 0 & 0 \\ i\varepsilon & 1 \end{pmatrix} \mathcal{K}(r + 1, s; \varepsilon),$$
(E.12)

which, with $r = x/\varepsilon$ and $s = t/\varepsilon$, produces the relation

$$\mathcal{K}\left(\frac{x}{\varepsilon}, \frac{(t + \varepsilon)}{\varepsilon}; \varepsilon\right) = \begin{pmatrix} 1 & i\varepsilon \\ 0 & 0 \end{pmatrix} \mathcal{K}\left(\frac{(x - \varepsilon)}{\varepsilon}, \frac{t}{\varepsilon}; \varepsilon\right)$$

$$+ \begin{pmatrix} 0 & 0 \\ i\varepsilon & 1 \end{pmatrix} \mathcal{K}\left(\frac{(x + \varepsilon)}{\varepsilon}, \frac{t}{\varepsilon}; \varepsilon\right). \quad \text{(E.13)}$$

[4] Compare with P. A. M. Dirac, "The Principles of Quantum Mechanics," pp. 260–262. Oxford Univ. Press, New York and London, 1947.

[5] By symmetry, we also have

$$N_{++}(r, s; R) = N_{--}(-r, s; R)$$

and

$$N_{+-}(r, s; R) = N_{-+}(r, s; R) = N_{+-}(-r, s; R) = N_{-+}(-r, s; R)$$

Combinatorial analysis shows that

$$N_{++}(r, s; R) = \{\tfrac{1}{2}(s + r - R) - 1, \tfrac{1}{2}R\}\{\tfrac{1}{2}(s - r - R), \tfrac{1}{2}R - 1\}$$

and

$$N_{+-}(r, s; R) = \{\tfrac{1}{2}(s + r - R) - \tfrac{1}{2}, \tfrac{1}{2}R - \tfrac{1}{2}\}\{\tfrac{1}{2}(s - r - R) - \tfrac{1}{2}, \tfrac{1}{2}R - \tfrac{1}{2}\}$$

for $R > 0$, where

$$\{m, n\} \equiv \begin{array}{ll} (m + n)!/m! \, n! & \text{for } m, n \text{ nonnegative integers,} \\ 0 & \text{otherwise.} \end{array}$$

An elegant derivation of these formulas for N_{++} and N_{+-} has been made by Charles Bock (unpublished).

By writing Taylor expansions for the terms in (E.13), we have

$$\mathcal{K}\left(\frac{x}{\varepsilon},\frac{t}{\varepsilon};\varepsilon\right) + \varepsilon\frac{\partial}{\partial t}\mathcal{K}\left(\frac{x}{\varepsilon},\frac{t}{\varepsilon};\varepsilon\right) + O(\varepsilon^2)\mathcal{K}\left(\frac{x}{\varepsilon},\frac{t}{\varepsilon};\varepsilon\right)$$

$$= \begin{pmatrix} 1 & i\varepsilon \\ 0 & 0 \end{pmatrix}\mathcal{K}\left(\frac{x}{\varepsilon},\frac{t}{\varepsilon};\varepsilon\right) - \begin{pmatrix} 1 & 0 \\ 0 & 0 \end{pmatrix}\varepsilon$$

$$\times\frac{\partial}{\partial x}\mathcal{K}\left(\frac{x}{\varepsilon},\frac{t}{\varepsilon};\varepsilon\right) + \begin{pmatrix} 0 & 0 \\ i\varepsilon & 1 \end{pmatrix}\mathcal{K}\left(\frac{x}{\varepsilon},\frac{t}{\varepsilon};\varepsilon\right)$$

$$+ \begin{pmatrix} 0 & 0 \\ 0 & 1 \end{pmatrix}\varepsilon\frac{\partial}{\partial x}\mathcal{K}\left(\frac{x}{\varepsilon},\frac{t}{\varepsilon};\varepsilon\right)$$

$$+ O(\varepsilon^2)\mathcal{K}\left(\frac{x}{\varepsilon},\frac{t}{\varepsilon};\varepsilon\right), \tag{E.14}$$

and so

$$\frac{\partial}{\partial t}\mathcal{K}\left(\frac{x}{\varepsilon},\frac{t}{\varepsilon};\varepsilon\right) = \begin{pmatrix} 0 & i \\ i & 0 \end{pmatrix}\mathcal{K}\left(\frac{x}{\varepsilon},\frac{t}{\varepsilon};\varepsilon\right)$$

$$- \begin{pmatrix} 1 & 0 \\ 0 & -1 \end{pmatrix}\frac{\partial}{\partial x}\mathcal{K}\left(\frac{x}{\varepsilon},\frac{t}{\varepsilon};\varepsilon\right) + O(\varepsilon)\mathcal{K}\left(\frac{x}{\varepsilon},\frac{t}{\varepsilon};\varepsilon\right). \tag{E.15}$$

The latter relation shows that (E.9) satisfies (E.5) with

$$\alpha = \begin{pmatrix} 1 & 0 \\ 0 & -1 \end{pmatrix}, \quad \text{and} \quad \beta = \begin{pmatrix} 0 & -1 \\ -1 & 0 \end{pmatrix}. \tag{E.16}$$

Moreover, from the definition of the N's and (E.9) it follows that

$$\lim_{t\to 0} K(x,t) = \lim_{\varepsilon\to 0}\mathcal{K}\left(\frac{x}{\varepsilon},1;\varepsilon\right)$$

$$= \lim_{\varepsilon\to 0}\frac{1}{\varepsilon}\begin{pmatrix} N_{++}\left(\frac{x}{\varepsilon},1;0\right) & N_{+-}\left(\frac{x}{\varepsilon},1;0\right) \\ N_{-+}\left(\frac{x}{\varepsilon},1;0\right) & N_{--}\left(\frac{x}{\varepsilon},1;0\right) \end{pmatrix}$$

$$= \begin{pmatrix} \delta(x) & 0 \\ 0 & \delta(x) \end{pmatrix}, \tag{E.17}$$

and thus (E.9) satisfies (E.6). Hence, the relativistic sum-over-histories (E.9) gives the propagation kernel for the one-dimensional Dirac equation. Unfortunately, a generalization of (E.9) for three spatial dimensions has not been found.

Appendix F Feynman Operators

Consider a generic functional $A = A[q(t)]$ that depends on $q(t)$ for $t' \leqslant t \leqslant t''$ and is analytic[1] about $q(t) \equiv 0$. Suppose further that the Feynman operator

$$\int_{\mathscr{G}} A(\exp\, iS/\hbar)D(\hat{q}) \equiv (q''; t'' \,|A[q(t)]|\, q'; t')_s, \qquad (\text{F.1})$$

exists for A. The Volterra expansion (A.19) for $A[q(t)]$ with $\bar{q}(t) \equiv 0$ takes the generic form

$$A[q(t)] = \sum_{k=0}^{\infty} \int_{t'}^{t''} \cdots \int_{t'}^{t''} \theta^{(k)}(t_1,\ldots, t_k)q(t_1) \cdots q(t_k)\, dt_1 \cdots dt_k, \qquad (\text{F.2})$$

where (suppressed) indices on the distribution-valued $\theta^{(k)}$'s and generalized coordinate n-tuple q's are understood to be contracted. Since the $\theta^{(k)}$'s are independent of $\hat{q}$, it follows from (F.1) that

$$(q''; t'' \,|A[q(t)]|\, q'; t')_s$$

$$= \sum_{k=0}^{\infty} \int_{t'}^{t''} \cdots \int_{t'}^{t''} \theta^{(k)}(t_1,\ldots, t_k)(q''; t'' \,|q(t_1) \cdots q(t_k)|\, q'; t')_s\, dt_1 \cdots dt_k.$$

$$(\text{F.3})$$

---

[1] From a practical point of view, there is no loss in generality in assuming that $A = A[q(t)]$ is analytic about $q(t) \equiv 0$. If not originally so, A can always be modified to be analytic about $q(t) \equiv 0$ without inducing a physically significant change in the associated Feynman operator.

Recalling the propagation kernel representations

$$K(q'', q'; t'' - t') = (q''; t'' |1| q'; t')_s$$

$$= (\exp -i\mathbf{H}''(t'' - t')/\hbar) \delta(q'' - q'), \qquad \text{(F.4)}$$

where $\mathbf{H}'' \equiv H(q'', -i\hbar \, \partial/\partial q'')$ is the Hermitian quantum Hamiltonian operator associated with the action functional S, we have

$$(q''; t'' |q(t_1) \cdots q(t_k)| q'; t')$$

$$\equiv \int_{\mathcal{G}} q(t_1) \cdots q(t_k)(\exp iS/\hbar)D(\hat{q}) = \int \cdots \int K(q'', q^{(k)}; t'' - t_{i_k})q^{(k)}$$

$$\times K(q^{(k)}, q^{(k-1)}; t_{i_k} - t_{i_{k-1}})q^{(k-1)} \cdots q^{(1)}K(q^{(1)}, q'; t_{i_1} - t') \, dq^{(1)}$$

$$\cdots dq^{(k)} = \int \cdots \int (\exp -i\mathbf{H}''(t'' - t_{i_k})/\hbar) \, \delta(q'' - q^{(k)})q^{(k)}$$

$$\times (\exp -i\mathbf{H}^{(k)}(t_{i_k} - t_{i_{k-1}})/\hbar) \, \delta(q^{(k)} - q^{(k-1)})q^{(k-1)} \cdots q^{(1)}$$

$$\times (\exp -i\mathbf{H}^{(1)}(t_{i_1} - t')/\hbar) \, \delta(\hat{q}^{(1)} - q') \, dq^{(1)} \cdots dq^{(k)}$$

$$= (\exp -i\mathbf{H}''t''/\hbar)\mathbf{q}''(t_{i_k})\mathbf{q}''(t_{i_{k-1}}) \cdots \mathbf{q}''(t_{i_1})$$

$$\times (\exp i\mathbf{H}''t'/\hbar) \, \delta(q'' - q') \qquad \text{(F.5)}$$

in which $t_{i_k} \geqslant t_{i_{k-1}} \geqslant \cdots \geqslant t_{i_1}$, and

$$\mathbf{q}''(t) \equiv (\exp i\mathbf{H}''t/\hbar)q''(\exp -i\mathbf{H}''t/\hbar). \qquad \text{(F.6)}$$

By substituting the final member of (F.5) into (F.3) and introducing the *chronological ordering symbol* T, which arranges noncommuting factors to the left with increasing values of t,

$$T(\mathbf{q}''(t_1) \cdots \mathbf{q}''(t_k)) \equiv \mathbf{q}''(t_{i_k}) \cdots \mathbf{q}''(t_{i_1}), \qquad t_{i_k} \geqslant \cdots \geqslant t_{i_1}, \quad \text{(F.7)}$$

we obtain the formula that relates Feynman operators to the more conventional differential operators of quantum mechanics[1],

$$(q''; t'' |A[q(t)]| q'; t')_s$$

$$= (\exp -i\mathbf{H}''t''/\hbar)T(A[\mathbf{q}''(t)])(\exp i\mathbf{H}''t'/\hbar) \, \delta(q'' - q'). \quad \text{(F.8)}$$

If time-derivatives of $q(t)$ appear in the functional $A[q(t)]$, the time-differentiations transfer to the $\theta^{(k)}$'s in (F.2), and it is clear that they must be put to the left of the T symbol; thus, for example,

$$T(\ddot{\mathbf{q}}(t_1)\mathbf{q}(t_2))$$

$$\equiv (d^2/dt_1{}^2)T(\mathbf{q}(t_1)\mathbf{q}(t_2))$$

$$= (d^2/dt_1{}^2)(\sigma(t_1 - t_2)\mathbf{q}(t_1)\mathbf{q}(t_2) + \sigma(t_2 - t_1)\mathbf{q}(t_2)\mathbf{q}(t_1))$$

$$= \sigma(t_1 - t_2)\ddot{\mathbf{q}}(t_1)\mathbf{q}(t_2) + \sigma(t_2 - t_1)\mathbf{q}(t_2)\ddot{\mathbf{q}}(t_1)$$

$$+ \delta(t_1 - t_2)(\dot{\mathbf{q}}(t_1)\mathbf{q}(t_2) - \mathbf{q}(t_2)\dot{\mathbf{q}}(t_1)) \tag{F.9}$$

in which

$$\sigma(u) \equiv \begin{array}{ll} 1 & \text{for} \quad u > 0 \\ \tfrac{1}{2} & \text{for} \quad u = 0 \\ 0 & \text{for} \quad u < 0. \end{array}$$

Equation (F.8) can be used to translate Feynman operator equations into a more conventional form that involves the familiar differential operators of Schrödinger and Heisenberg. For example, the operator Euler–Lagrange equations

$$(q''; t'' |\delta S/\delta q(t)| q'; t')_s = 0 \tag{F.10}$$

produce

$$\ddot{\mathbf{q}}(t) + \partial V(q)/\partial q \bigg|_{q = \mathbf{q}(t)} = 0 \tag{F.11}$$

for the action functional

$$S = \int_{t'}^{t''} \left(\tfrac{1}{2}\, \dot{q} \cdot \dot{q} - V(q) \right) dt. \tag{F.12}$$

The Feynman operator equations

$$(q''; t'' |[\delta S/\delta q_i(t_1)]q_j(t_2)| q'; t')_s = i\hbar\, \delta_{ij}\, \delta(t_1 - t_2)(q''; t'' |1| q'; t')_s \tag{F.13}$$

produce

$$T\left(\ddot{\mathbf{q}}_i(t_1)\mathbf{q}_j(t_2) + \partial V(q)/\partial q_i \bigg|_{q=\mathbf{q}(t_1)} \mathbf{q}_j(t_2) \right) = -i\hbar\, \delta_{ij}\, \delta(t_1 - t_2), \tag{F.14}$$

where the identity operator is understood to appear on the right-hand side; by making use of (F.9), the operator equations (F.13) become

$$\sigma(t_1 - t_2)\left(\ddot{\mathbf{q}}_i(t_1) + \partial V(q)/\partial q_i\Big|_{q=\mathbf{q}(t_1)}\right)\mathbf{q}_j(t_2)$$

$$+ \sigma(t_2 - t_1)\mathbf{q}_j(t_2)\left(\ddot{\mathbf{q}}_i(t_1) + \partial V(q)/\partial q_i\Big|_{q=\mathbf{q}(t_1)}\right)$$

$$+ \delta(t_1 - t_2)(\dot{\mathbf{q}}_i(t_1)\mathbf{q}_j(t_2) - \mathbf{q}_j(t_2)\dot{\mathbf{q}}_i(t_1)) = -i\hbar\, \delta_{ij}\, \delta(t_1 - t_2) \quad \text{(F.15)}$$

and reduce to the Heisenberg commutation relations

$$\dot{\mathbf{q}}_i(t)\mathbf{q}_j(t) - \mathbf{q}_j(t)\dot{\mathbf{q}}_i(t) = -i\hbar\, \delta_{ij} \tag{F.16}$$

because of the operator Euler–Lagrange equations (F.11). Finally, we note that in the case of the action functional (F.12) with

$$V(q) = \tfrac{1}{2}\sum_{i=1}^{n} \omega_i^2 q_i^2, \tag{F.17}$$

the Feynman operator equations (F.13) take a form which is translated by (F.8) into the operator equations

$$(d^2/dt_1^2 + \omega_i^2)T(\mathbf{q}_i(t_1)\mathbf{q}_j(t_2)) = -i\hbar\, \delta_{ij}\, \delta(t_1 - t_2). \tag{F.18}$$

Equations (F.18) enable one to determine the *causal Green's function* of perturbation theory, the ground state expectation value of $T(\mathbf{q}_i(t_1)\mathbf{q}_j(t_2))$.[2]

[2] R. P. Feynman and A. R. Hibbs, "Quantum Mechanics and Path Integrals," p. 180. McGraw-Hill, New York, 1965.

Appendix G *Quantum and Classical Statistical Mechanics*

Consider a dynamical system described by a Hamiltonian of the form $H = U + V$, where the kinetic $U = U(p)$ is a function of a generalized linear momentum n-tuple and the potential energy $V = V(q)$ is a function of a generalized linear coordinate n-tuple. For the quantum statistical description of the system, we have a (microcanonical) ensemble of normalized wave functions $\psi_\mu = \psi_\mu(q; t)$ that satisfy the Schrödinger equation

$$i\hbar \, \partial\psi_\mu/\partial t = [U(-i\hbar \, \partial/\partial q) + V(q)]\psi_\mu \qquad (G.1)$$

with μ an enumerator index. Expectation values of observables are defined by[1]

$$\langle f \rangle_\hbar \equiv \sum_\mu \omega_\mu \int \psi_\mu^*(q; t) f(q, -i\hbar \, \partial/\partial q)\psi_\mu(q; t) \, dq \qquad (G.2)$$

with the canonical variables in $f(q, p)$ ordered to give an Hermitian operator.[2] The ω_μ's in (G.2) are probability weights for the states in

---

[1] In order to have an enumerable set of states for H which admit running waves, so-called box normalization is understood here.

[2] Our formalism does not require a specification of the Hermitian ordering rule (or rules); instead, the ordering of quantum observables is essentially free to be determined appropriately by experiment.

the ensemble, positive constants that sum to unity,

$$\omega_\mu > 0, \qquad \sum_\mu \omega_\mu = 1. \tag{G.3}$$

Let us introduce the *hypercharacteristic function*[3]

$$\Phi_\hbar = \Phi_\hbar(\xi, \eta; t) \equiv \sum_\mu \omega_\mu \int e^{i\xi \cdot q} \psi_\mu^*(q - \tfrac{1}{2}\hbar\eta; t)\psi_\mu(q + \tfrac{1}{2}\hbar\eta; t)\, dq \tag{G.4}$$

with the properties[4]

$$\Phi^*_\hbar(\xi, \eta; t) = \Phi_\hbar(-\xi, -\eta; t), \qquad \Phi_\hbar(0, 0; t) = 1. \tag{G.5}$$

Expectation values are obtained directly from the hypercharacteristic function according to the formula

$$\langle f \rangle_\hbar = [f(-i\,\partial/\partial\xi + \tfrac{1}{2}\hbar\eta, -i\,\partial/\partial\eta - \tfrac{1}{2}\hbar\xi)\Phi_\hbar]_{\xi = 0 = \eta}. \tag{G.6}$$

To prove (G.6), we first note the operator relations

$$(-i\,\partial/\partial\xi + \tfrac{1}{2}\hbar\eta) \exp[i\xi \cdot (q + \tfrac{1}{2}\hbar\eta)]$$
$$= \exp[i\xi \cdot (q + \tfrac{1}{2}\hbar\eta)](-i\,\partial/\partial\xi + q + \hbar\eta)$$
$$(-i\,\partial/\partial\eta - \tfrac{1}{2}\hbar\xi) \exp[i\xi \cdot (q + \tfrac{1}{2}\hbar\eta)]$$
$$= \exp[i\xi \cdot (q + \tfrac{1}{2}\hbar\eta)](-i\,\partial/\partial\eta), \tag{G.7}$$

from which it follows that for a generic $f(q, p)$ with the arguments ordered in a definite way,[2] we have the operator relation

$$f(-i\,\partial/\partial\xi + \tfrac{1}{2}\hbar\eta, -i\,\partial/\partial\eta - \tfrac{1}{2}\hbar\xi)\exp[i\xi \cdot (q + \tfrac{1}{2}\hbar\eta)]$$
$$= \exp[i\xi \cdot (q + \tfrac{1}{2}\hbar\eta)]f(-i\,\partial/\partial\xi + q + \hbar\eta, -i\,\partial/\partial\eta). \tag{G.8}$$

[3] This function is the double Fourier transform of the *quasiprobability density* defined by E. Wigner, *Phys. Rev.* **40**, 749 (1932). For a *pure state* (that is, one value of μ entering the summations in (G.2) and (G.4) with the ω_μ equal to unity) *wave packet*, the quasiprobability density

$$(2\pi)^{-n} \int e^{-i\eta \cdot p} \psi^*(q - \tfrac{1}{2}\hbar\eta; t)\psi(q + \tfrac{1}{2}\hbar\eta; t)\, d\eta$$

is significant in magnitude through a q, p phase space volume about equal to $(2\pi\hbar)^n$ and displays a mean motion along the classical trajectory in phase space.

[4] The hypercharacteristic function also possesses nonholonomic properties, as exemplified by the bounding condition $|\Phi_\hbar| \leqslant 1$; other nonholonomic properties follow from the results of G. A. Baker, Jr., *Phys. Rev.* **109**, 2198 (1958); T. Takabayasi, *Progr. Theor. Phys.* **11**, 341 (1954).

Next, we let both sides of (G.8) act on $\psi_\mu(q + \hbar\eta; t)$, multiply from the left by $\omega_\mu\psi_\mu{}^*(q; t)$, integrate over q and sum over μ to get

$$f(-i\,\partial/\partial\xi + \tfrac{1}{2}\hbar\eta, -i\,\partial/\partial\eta - \tfrac{1}{2}\hbar\xi)\Phi_\hbar$$

$$= \sum_\mu \omega_\mu \int \psi_\mu{}^*(q; t)\exp[i\xi \cdot (q + \tfrac{1}{2}\hbar\eta)]$$

$$\times [f(-i\,\partial/\partial\xi + q + \hbar\eta, -i\,\partial/\partial\eta)\psi_\mu(q + \hbar\eta; t)]\,dq, \qquad \text{(G.9)}$$

where use is made of the definition (G.4) with a displacement of the integration variable $q \to (q + \tfrac{1}{2}\hbar\eta)$. The square bracket term in (G.9) equals

$$[f(-i\,\partial/\partial\xi + q + \hbar\eta, -i\hbar\,\partial/\partial q)\psi_\mu(q + \hbar\eta; t)]$$

$$= [f(q + \hbar\eta, -i\hbar\,\partial/\partial q)\psi_\mu(q + \hbar\eta; t)], \qquad \text{(G.10)}$$

independent of how the arguments are ordered in $f(q, p)$, and so from (G.9) and (G.2) we obtain formula (G.6). In view of the Schrödinger equation (G.1), the dynamical evolution of the hypercharacteristic function (and thus of expectation values (G.6)) is expressed by

$$\partial\Phi_\hbar/\partial t = (i\hbar[U(-i\,\partial/\partial\eta + \tfrac{1}{2}\hbar\xi) - U(-i\,\partial/\partial\eta - \tfrac{1}{2}\hbar\xi)$$

$$+ V(-i\,\partial/\partial\xi - \tfrac{1}{2}\hbar\eta) - V(-i\,\partial/\partial\xi + \tfrac{1}{2}\hbar\eta)]\Phi_\hbar. \qquad \text{(G.11)}$$

Essentially time-dependent problems in quantum statistical mechanics can be solved by integrating (G.11) subject to a prescribed initial form $\Phi_\hbar(\xi, \eta; 0)$, but equilibrium stationary states are not determined completely by the static solutions to (G.11), requiring an additional postulate for the theoretical determination of statistical equilibrium. With no condition imposed on the value of p, the probability density for values of q is

$$P_\hbar(q; t) \equiv \sum_\mu \omega_\mu |\psi_\mu(q; t)|^2 = (2\pi)^{-n} \int e^{-i\xi \cdot q}\Phi_\hbar(\xi, 0; t)\,d\xi.$$

$$\text{(G.12)}$$

The formalism for classical statistical mechanics appears as the classical limit, $\hbar \to 0$, of the preceding formalism for quantum statistical mechanics. Expectation values in classical statistical mechanics are identified with the classical limit of the expectation values in quantum statistical mechanics given by (G.6),

$$\langle f \rangle_0 \equiv \lim_{\hbar \to 0} \langle f \rangle_\hbar = [f(-i\, \partial/\partial\xi, \, -i\, \partial/\partial\eta)\Phi_0]_{\xi=0=\eta}, \quad \text{(G.13)}$$

in which the *characteristic function* of classical statistical mechanics,

$$\Phi_0 = \Phi_0(\xi, \eta; t) \equiv \lim_{\hbar \to 0} \Phi_\hbar(\xi, \eta; t), \quad \text{(G.14)}$$

satisfies the limiting form of (G.11),[5]

$$\partial\Phi_0/\partial t = i[\xi \cdot \text{grad } U(-i\, \partial/\partial\eta) - \eta \cdot \text{grad } V(-i\, \partial/\partial\xi)]\Phi_0, \quad \text{(G.15)}$$

where *n*-tuple gradients of the energy terms are denoted by grad $U(p) \equiv \partial U/\partial p$ and grad $V(q) \equiv \partial V/\partial q$. The probability density for values of q in classical statistical mechanics, with no condition imposed on the value of p, is identified with the classical limit of (G.12),

$$P_0(q; t) \equiv \lim_{\hbar \to 0} P_\hbar(q; t) = (2\pi)^{-n} \int e^{-i\xi \cdot q} \Phi_0(\xi, 0; t)\, d\xi. \quad \text{(G.16)}$$

We introduce the Fourier transform of Φ_0,

$$W = W(q, p; t) \equiv (2\pi)^{-2n} \int e^{-i\xi \cdot q - i\eta \cdot p} \Phi_0(\xi, \eta; t)\, d\xi\, d\eta, \quad \text{(G.17)}$$

a real normalized function because of the properties of the hyper-characteristic function (G.5),

$$W(q, p; t)^* = W(q, p; t), \qquad \int W(q, p; t)\, dq\, dp = 1. \quad \text{(G.18)}$$

Interpreted as the phase space *probability density* for values of q and p over the (microcanonical) ensemble, it can be shown that the quantity (G.17) is nonnegative as a consequence of definitions (G.4) and (G.14).[4] In terms of the probability density (G.17), Eqs. (G.13), (G.15), and (G.16) reduce to the familiar form for classical statistical mechanics:

$$\langle f \rangle_0 = \int f(q, p)W(q, p; t)\, dq\, dp, \quad \text{(G.19)}$$

$$\partial W/\partial t = -(\text{grad } U(p)) \cdot \partial W/\partial q + (\text{grad } V(q)) \cdot \partial W/\partial p = [H, W], \quad \text{(G.20)}$$

$$P_0(q; t) = \int W(q, p; t)\, dp. \quad \text{(G.21)}$$

[5] Note that for $U(p)$ quadratic in p and $V(q)$ quadratic in q (that is, for H an element of the Lie algebra $\mathscr{A}_s$), $\hbar$ cancels out in the differential operator on the right-hand side of (G.11), and the dynamical equations for $\Phi_\hbar$ and Φ_0 are identical (reflecting isomorphism of the quantum and classical Lie algebras $\mathscr{A}_s$ and $\mathscr{A}_s$).

Appendix H   *The Iteration Solution of Perturbation Theory*

Consider a dynamical system described by a Hamiltonian of the form $H = H_0 + H_{int}$, where $H_0 = H_0(q, p)$ is simple (that is, quadratic in the canonical variables) and $H_{int} = H_{int}(q)$ is relatively small compared to H_0 for typical values of the canonical variables. The Schrödinger quantum dynamical description of the system is based on the equation

$$i\hbar\, \partial\psi/\partial t = (\mathbf{H}_0 + H_{int})\psi \tag{H.1}$$

satisfied by the wave function $\psi = \psi(q; t)$, where

$$\mathbf{H}_0 \equiv H_0(q,\, -i\hbar\, \partial/\partial q). \tag{H.2}$$

It follows from (H.1) that the *interaction wave function*

$$\psi_{int} = \psi_{int}(q; t) \equiv (\exp i\mathbf{H}_0 t/\hbar)\psi \tag{H.3}$$

satisfies the equation

$$i\hbar\, \partial\psi_{int}/\partial t = H_{int}(\mathbf{q}(t))\psi_{int}, \tag{H.4}$$

with

$$\mathbf{q}(t) \equiv (\exp i\mathbf{H}_0 t/\hbar)q(\exp -i\mathbf{H}_0 t/\hbar)$$

so that

$$H_{int}(\mathbf{q}(t)) = (\exp i\mathbf{H}_0 t/\hbar)H_{int}(q)(\exp -i\mathbf{H}_0 t/\hbar). \tag{H.5}$$

131

Subject to a prescribed form for the interaction wave function at $t = t'$, the formal solution to (H.4) at a subsequent $t = t''$ is given by

$$\psi_{int}(q; t'') = T\left(\exp -i \int_{t'}^{t''} H_{int}(\mathbf{q}(t))\, dt/\hbar\right)\psi_{int}(q; t'), \quad (H.6)$$

in which the *chronological ordering symbol T* arranges noncommuting factors to the left with increasing values of t,

$$T(H_{int}(\mathbf{q}(t_1)) \cdots H_{int}(\mathbf{q}(t_k)))$$
$$\equiv H_{int}(\mathbf{q}(t_{i_k})) \cdots H_{int}(\mathbf{q}(t_{i_1})), \quad t_{i_k} \geqslant \cdots \geqslant t_{i_1}. \quad (H.7)$$

We verify that (H.6) satisfies (H.4) by noting the relation

$$T\left(\exp -i \int_{t'}^{t''+\varepsilon} H_{int}(\mathbf{q}(t))\, dt/\hbar\right)$$
$$= (1 - i\varepsilon H_{int}(\mathbf{q}(t''))/\hbar + O(\varepsilon^2))T\left(\exp -i \int_{t'}^{t''} H_{int}(\mathbf{q}(t))\, dt/\hbar\right), \quad (H.8)$$

which implies that

$$\left(i\hbar \frac{\partial}{\partial t''} - H_{int}(\mathbf{q}(t''))\right)T\left(\exp -i \int_{t'}^{t''} H_{int}(\mathbf{q}(t))\, dt/\hbar\right) = 0. \quad (H.9)$$

The *iteration solution*[1] of perturbation theory is obtained by expanding the exponential in (H.6),

$$\psi_{int}(q; t'') = \sum_{k=0}^{\infty} (k!)^{-1}(i/\hbar)^k T\left(- \int_{t'}^{t''} H_{int}(\mathbf{q}(t))\, dt\right)^k \psi_{int}(q; t'). \quad (H.10)$$

Moreover, by inverting the definition (H.3) and recalling the propagation kernel equation

$$\psi(q''; t'') = \int K(q'', q'; t'' - t')\psi(q'; t')\, dq', \quad (H.11)$$

we obtain the formula

$$K(q'', q'; t'' - t')$$
$$= \sum_{k=0}^{\infty} (k!)^{-1}(i/\hbar)^k (\exp -i\mathbf{H}_0''t''/\hbar)$$
$$\times T\left(- \int_{t'}^{t''} H_{int}(\mathbf{q}''(t))\, dt\right)^k (\exp i\mathbf{H}_0''t'/\hbar)$$
$$\times \delta(q'' - q') \quad (H.12)$$

[1] F. J. Dyson, *Phys. Rev.* **75**, 486 (1949).

where

$$\mathbf{H_0}'' \equiv H_0(q'', -i\hbar \, \partial/\partial q'')$$

$$\mathbf{q}''(t) \equiv (\exp i\mathbf{H_0}''t/\hbar)q''(\exp -i\mathbf{H_0}''t/\hbar). \tag{H.13}$$

According to the formula (F.8), the terms in (H.12) are Feynman operators computed with respect to the main part of the action functional $S = S_0 + S_{int}$ associated with $H = H_0 + H_{int}$,

$$S_0 = S_0[q(t)] = \int_{t'}^{t''} L_0(q(t), \dot{q}(t)) \, dt, \tag{H.14}$$

where $L_0(q, \dot{q}) = p \cdot \dot{q} - H_0(q, p)$ with p eliminated in favor of q and $\dot{q}$ by evoking the n-tuple Hamilton canonical equation $\dot{q} = \partial H/\partial p = \partial H_0/\partial p$. In view of (H.12) and (F.8), we have[2]

$$K(q'', q'; t'' - t') = \sum_{k=0}^{\infty} (k!)^{-1}(i/\hbar)^k (q''; t'' |(S_{int})^k| q'; t')_{so}$$

$$= (q''; t''|(\exp iS_{int}/\hbar)| q'; t')_{so}$$

$$= (q''; t'' |1| q'; t')_s, \tag{H.15}$$

where

$$S_{int} \equiv S - S_0 = S_{int}[q(t)] = -\int_{t'}^{t''} H_{int}(q(t)) \, dt. \tag{H.16}$$

Conversely, the standard form of the interation solution (H.10) follows deductively from (H.15).

---

[2] Direct applications of (H.15) are worked out by R. P. Feynman and A. R. Hibbs, "Quantum Mechanics and Path Integrals," pp. 120–161. McGraw-Hill, New York, 1965.

Appendix I Existence of Spatially Localized Singularity-Free Periodic Solutions in Classical Field Theories

A necessary condition for the existence of spatially localized singularity-free periodic solutions to Euler–Lagrange classical field equations is readily obtainable by applying dilatation (scale) considerations to the variables in the action principle. In particular, the dilatation considerations can be applied to the action principles for Einstein-type unified theories,[1] in which the existence of spatially localized singularity-free periodic solutions is crucial to the physical interpretation.

We illustrate the dilatation considerations that lead to the necessary condition for existence of spatially localized singularity-free periodic solutions by treating the generic local theory based on the Lagrangian density

$$\mathcal{L} = \tfrac{1}{2}\dot{\phi} \cdot \dot{\phi} - \tfrac{1}{2}\sum_{i=1}^{3} \nabla_i \phi \cdot \nabla_i \phi - u(\phi), \tag{I.1}$$

where $\phi = \phi(x; t)$ is a field amplitude real n-tuple and $u(\phi)$ is a self-interaction energy density real 1-tuple. The action principle

$$\delta S/\delta\phi(x; t) = 0 \qquad S = S[\phi(x; t)] \equiv \int_0^T \int \mathcal{L} \, dx \, dt \tag{I.2}$$

---

[1] A. Einstein, "The Meaning of Relativity," pp. 139–166. Princeton Univ. Press, Princeton, New Jersey, 1955.

134

for a solution periodic in time, $\phi(x; t + T) = \phi(x; t)$, implies the global condition

$$(dS[\phi(\lambda x; t)]/d\lambda)_{\lambda-1} = 0 \qquad (I.3)$$

for a solution localized in space, where λ is a real positive parameter associated with the specific variation of the field amplitude $\phi(x; t) \rightarrow \phi(\lambda x; t) \equiv \phi(x; t) + \delta\phi(x; t)$. Provided that the x integrations converge, we have

$$\int \dot{\phi}(\lambda x; t) \cdot \dot{\phi}(\lambda x; t) \, dx = \lambda^{-3} \int \dot{\phi}(x; t) \cdot \dot{\phi}(x; t) \, dx \qquad (I.4)$$

and similar relations for the other terms in $\int \mathcal{L} \, dx$ with $\phi = \phi(\lambda x; t)$; thus, (I.3) works out to give

$$\int_0^T \int \left(-\tfrac{3}{2}\dot{\phi} \cdot \dot{\phi} + \tfrac{1}{2}\sum_{i=1}^3 \nabla_i \phi \cdot \nabla_i \phi + 3u(\phi) \right) dx \, dt = 0 \qquad (I.5)$$

with $\phi = \phi(x; t)$. An alternative way to derive (I.5) is to contract the field equations prescribed by (I.1) and (I.2),

$$\ddot{\phi} - \nabla^2\phi + \partial u(\phi)/\partial\phi = 0, \qquad (I.6)$$

with the real n-tuple functions $\left(\sum_{i=1}^3 x_i \nabla_i \phi\right)$ and integrate the resulting equation over all x and over the range 0 to T for t. On the other hand, if we work out

$$(dS[\lambda^{-1}\phi(x; t)]/d\lambda)_{\lambda=1} = 0, \qquad (I.7)$$

or if we contract (I.6) with ϕ and integrate, we obtain

$$\int_0^T \int \left(-\dot{\phi} \cdot \dot{\phi} + \sum_{i=1}^3 \nabla_i \phi \cdot \nabla_i \phi + \phi \cdot (\partial u(\phi)/\partial\phi) \right) dx \, dt = 0, \qquad (I.8)$$

again provided that the x integrations converge and T is the period of the solution. Finally, by subtracting two-thirds of (I.5) from (I.8), we find the global condition

$$\int_0^T \int \left(\tfrac{2}{3}\sum_{i=1}^3 \nabla_i \phi \cdot \nabla_i \phi + \phi \cdot (\partial u(\phi)/\partial\phi) - 2u(\phi) \right) dx \, dt = 0, \qquad (I.9)$$

which implies that

$$\tfrac{1}{2}\phi \cdot (\partial u(\phi)/\partial\phi) < u(\phi) \qquad (I.10)$$

for some values of ϕ is a necessary condition for existence of a localized singularity-free periodic solution to Eqs. (I.6). Hence, such solutions are precluded in theories with positive-definite self-interaction energy densities of simple algebraic form, as exemplified by

$$u(\phi) = m^2\phi \cdot \phi + g(\phi \cdot \phi)^2 \tag{I.11}$$

with m^2, g real positive constants. Spatially localized singularity-free periodic solutions have been found, however, for local field theories with Lagrangian densities of the form (I.1) that feature more complicated positive-definite self-interaction energy densities (satisfying the condition (I.10)).[2]

[2] G. Rosen, *J. Math. Phys.* **9**, 966 (1968), and works cited therein.

Appendix J **Rigorous Solutions in Essentially Nonlinear Classical Field Theories**

Exact solutions to the field equations of an essentially nonlinear theory follow from the reduction of the system of partial differential equations in the space-time independent variables to a system of ordinary differential equations in one independent variable, there being a well-developed general integration theory for systems of non-linear *ordinary* differential equations.[1] Ignoring trivial reductions in which the field is prescribed to depend only on a single space-time coordinate, one must apply a group-theoretic method[2,3] (based on the invariance character of the specific field equations) in order to accomplish the reduction to a system of ordinary differential equations.

We illustrate the group-theoretic method of reduction by considering classical field equations of the form

$$-\ddot{\phi} + \nabla^2\phi \equiv \partial^2\phi/\partial x^\mu \, \partial x_\mu = \partial u(\phi)/\partial\phi, \qquad (J.1)$$

---

[1] For example, A. R. Forsyth, "Theory of Differential Equations," Vols. II and III. Dover, New York, 1959.

[2] A. J. A. Morgan, *Quart. J. Math., Oxford Ser.* **2**, 250 (1952).

[3] G. Birkhoff, "Hydrodynamics," Chaps 4 and 5. Princeton Univ. Press, Princeton, New Jersey, 1960.

where the components of the real n-tuple field amplitude ϕ are scalar functions of the space-time coordinates $t \equiv x^0 \equiv -x_0$, $x \equiv (x_1, x_2, x_3) \equiv (x^1, x^2, x^3)$, and $u(\phi)$ is a self-interaction energy density real 1-tuple. Manifest Lorentz-invariance of the n field equations (J.1) implies[2] the existence of solutions of the form

$$\phi = \alpha(r), \tag{J.2}$$

in which the Lorentz-invariant scalar argument of α is

$$r \equiv \tfrac{1}{4} x^\mu x_\mu \equiv \tfrac{1}{4}(-t^2 + x_1{}^2 + x_2{}^2 + x_3{}^2). \tag{J.3}$$

Indeed, by putting (J.2) into (J.1), we obtain the system of ordinary differential equations for $\alpha = \alpha(r)$,

$$r(d^2\alpha/dr^2) + 2(d\alpha/dr) = \partial u(\alpha)/\partial\alpha. \tag{J.4}$$

Less obvious reductions of Eqs. (J.1) to systems of ordinary differential equations depend on the specific algebraic character of the self-interaction energy density. For $u(\phi)$ a homogeneous function of the field amplitude n-tuple,

$$u(\lambda\phi) \equiv \lambda^\gamma u(\phi) \qquad \text{for all real} \quad \lambda > 0, \tag{J.5}$$

where γ is the degree of homogeneity, the group-theoretic method of reduction is based on the invariance of Eqs. (J.1) under space-time dilatation transformations

$$x^\mu \to \lambda^{-1} x^\mu, \qquad \phi \to \lambda^{2/(\gamma-2)}\phi. \tag{J.6}$$

From the invariance under dilatation transformations (J.6), it follows that if $\phi(x^\mu)$ is a solution to (J.1), then so is $\lambda^{2/(\gamma-2)}\phi(\lambda x^\mu)$ for all real $\lambda > 0$. Moreover, a general theorem[2] guarantees existence of *self-similar solutions* to (J.1) such that

$$\phi(x^\mu) \equiv \lambda^{2/(\gamma-2)}\phi(\lambda x^\mu) \qquad \text{for all real} \quad \lambda > 0 \tag{J.7}$$

if the self-interaction energy density has the homogeneity prescribed by (J.5). Furthermore, it is possible to reduce Eqs. (J.1) to systems of ordinary differential equations for specific self-similar solutions.[3] For example, if we seek a self-similar solution that depends on $r(>0)$ defined by (J.3) and $(k_\mu x^\mu)$ with the k's constants, it must take the form

$$\phi = r^{1/(2-\gamma)}\beta(s), \qquad [\gamma \neq 2], \tag{J.8}$$

where the argument of β is the dilatation-invariant scalar

$$s \equiv (k_\mu x^\mu)/r^{1/2}. \tag{J.9}$$

Indeed, by putting (J.8) into (J.1), terms in powers of r cancel out, and we obtain the system of ordinary differential equations for $\beta = \beta(s)$,

$$(k_\mu k^\mu - \tfrac{1}{4}s^2)\frac{d^2\beta}{ds^2} - \tfrac{3}{4}s\frac{d\beta}{ds} + \frac{(3-\gamma)}{(2-\gamma)^2}\beta = \frac{\partial u(\beta)}{\partial \beta}. \tag{J.10}$$

If alternative self-similar forms are taken in place of the *ansatz* (J.8), the systems of ordinary differential equations that result admit determination of other rigorous solutions to Eqs. (J.1) with (J.5).

Appendix K *Quantum Theory of Electromagnetic Radiation*

Let us consider the electromagnetic field associated with a prescribed static solenoidal current density $j = j(x) = (j_1(x), j_2(x), j_3(x))$, $\partial j(x)/\partial t \equiv 0 \equiv \nabla \cdot j(x)$, and a vanishing electric charge density. Such a field is a linear superposition of free radiation and magnetostatic parts. Letting $\phi = \phi(x) = (\phi_1(x), \phi_2(x), \phi_3(x))$ denote the electromagnetic vector potential and $\pi = \pi(x) = (\pi_1(x), \pi_2(x), \pi_3(x))$ the canonically conjugate electric vector field, the Hamiltonian is given by

$$H = H[\phi, \pi] = \int (\tfrac{1}{2}\pi \cdot \pi + \tfrac{1}{2}(\text{curl } \phi) \cdot (\text{curl } \phi) + \phi \cdot j)\, dx. \qquad (K.1)$$

In order to complete the dynamical description for the electromagnetic radiation, the Hamilton canonical field equations derived from (K.1),

$$\dot{\phi}(x) = \delta H/\delta \pi(x) = \pi(x) \qquad (K.2)$$

$$\dot{\pi}(x) = -\delta H/\delta \phi(x) = -\text{curl}(\text{curl } \phi(x)) - j(x), \qquad (K.3)$$

must be supplemented by the physical condition on the electric vector field for a vanishing charge density,

$$\nabla \cdot \pi(x) = 0. \qquad (K.4)$$

140

Equations (K.2) and (K.3) combine to yield an inhomogeneous linear wave equation for the vector potential,

$$\ddot{\phi}(x) + \text{curl}(\text{curl }\phi(x)) = -j(x). \tag{K.5}$$

By introducing the transverse part of the vector potential

$$\phi_i^{\text{tr}}(x) \equiv \int \delta_{ij}^{\text{tr}}(x - y)\phi_j(y)\,dy \equiv \phi_i(x) - \nabla^{-2}\,\partial^2\phi_j(x)/\partial x_i\,\partial x_j, \tag{K.6}$$

in which the transverse δ-function has the Fourier representation

$$\delta_{ij}^{\text{tr}}(x) = (2\pi)^{-3} \int \{\delta_{ij} - [k_i k_j/(k \cdot k)]\}(\exp ik \cdot x)\,dk, \tag{K.7}$$

we find that Eq. (K.5) splits up to give

$$\ddot{\phi}^{\text{tr}}(x) - \nabla^2\phi^{\text{tr}}(x) = -j(x) \tag{K.8}$$

$$\nabla \cdot \ddot{\phi}(x) = 0. \tag{K.9}$$

It should be noted that the consistency of (K.4) for all time is guaranteed by (K.2) and the latter dynamical equation (K.9).

To effect quantization of the electromagnetic radiation field theory, we satisfy the canonical variable commutation relations

$$[\phi_i(x; t), \pi_j(y; t)] \equiv (i\hbar)^{-1}(\phi_i(x; t)\pi_j(y; t) - \pi_j(y; t)\phi_i(x; t))$$
$$= \delta_{ij}\,\delta(x - y) \tag{K.10}$$

at the instant of time $t = 0$ by taking the vector potential to be diagonal for all x, $\phi(x; 0) \equiv \phi(x)$, and representing the electric vector field by means of a functional differential operator,

$$\pi_i(x; 0) = -i\hbar\,\delta/\delta\phi_i(x). \tag{K.11}$$

Then from (K.1), we obtain the quantum Hamiltonian

$$\mathbf{H} = H[\phi(x; 0), \pi(x; 0)]$$

$$= \int \left(-\frac{\hbar^2}{2}\frac{\delta}{\delta\phi(x)} \cdot \frac{\delta}{\delta\phi(x)} + \frac{1}{2}\phi^{\text{tr}}(x) \cdot (-\nabla^2)\phi^{\text{tr}}(x) + \phi^{\text{tr}}(x) \cdot j(x) \right) dx,$$

$$\tag{K.12}$$

where the transverse part of the vector potential (K.6) is brought in through the relations curl $\phi(x) \equiv$ curl $\phi^{\text{tr}}(x)$ and $\nabla \cdot j(x) \equiv 0$. In view of (K.4) and (K.11), physically admissible wave functionals $\Psi[\phi; t]$ must satisfy the condition

$$\nabla \cdot \delta\Psi[\phi; t]/\delta\phi(x) = 0. \tag{K.13}$$

Hence, physically admissible wave functionals depend only on the transverse part of the vector potential,

$$\Psi[\phi; t] \equiv \Psi[\phi^{tr}; t], \tag{K.14}$$

because (K.13) implies

$$\frac{\delta\Psi[\phi + \nabla\chi; t]}{\delta\chi(x)} = -\nabla \cdot \frac{\delta\Psi[\phi + \nabla\chi; t]}{\delta\phi(x)} \equiv 0$$

for arbitrary real 1-tuple functions $\chi = \chi(x)$, and shows that

$$\Psi[\phi + \nabla\chi; t]$$

is independent of χ.

Stationary states of the field are represented by wave functionals $\Psi[\phi; t] = (\exp -iE_\mu t/\hbar)U_\mu[\phi]$ that satisfy the Schrödinger equation. Thus, we have

$$\mathbf{H}U_\mu[\phi] = E_\mu U_\mu[\phi], \tag{K.15}$$

where (K.14) requires $U_\mu[\phi]$ to depend only on the transverse part of the vector potential,

$$U_\mu[\phi] \equiv U_\mu[\phi^{tr}]. \tag{K.16}$$

Subject to (K.16), the ground state eigenfunctional solution to (K.15) with the functional differential operator quantum Hamiltonian (K.12) is given to within normalization by

$$U_0[\phi] = \exp -\hbar^{-1} \int (\tfrac{1}{2}\phi^{tr} \cdot (-\nabla^2)^{1/2}\phi^{tr} + \phi^{tr} \cdot (-\nabla^2)^{-1/2}j) \, dx,$$

$$\tag{K.17}$$

and the associated ground state energy eigenvalue is

$$E_0 = \hbar[(-\nabla^2)^{1/2} \, \delta(x)]_{x=0} \int dx - \int \tfrac{1}{2}j \cdot (-\nabla^2)^{-1}j \, dx. \tag{K.18}$$

It is only necessary to work out the second functional derivatives of (K.17) and to recall the definition

$$\delta^2 U_0/\delta\phi_i(x)^2 = \lim_{y \to x} \delta^2 U_0/\delta\phi_i(x) \, \delta\phi_i(y)$$

in order to verify that (K.17), (K.18) satisfy (K.15) with (K.12). The first term in the ground state energy (K.18) is the zero-point energy of the electromagnetic radiation vacuum, an unobservable infinite constant, while the second term in (K.18) is the classical Ampère self-interaction energy of the current density,[1] expressed in the familiar form

$$-\int [j(x) \cdot j(y)/8\pi |x-y|] dx\, dy$$

by using the relation $(-\nabla^2)^{-1} \delta(x - y) = (4\pi|x - y|)^{-1}$. To obtain the excited state eigenfunctional solutions to (K.15) with $E_\mu > E_0$, it is convenient to introduce the quantity $\Omega_\mu = \Omega_\mu[\phi] \equiv U_\mu[\phi]/U_0[\phi] \equiv \Omega_\mu[\phi^{tr}]$ which satisfies the simpler functional differential equation

$$\int \sum_{i=1}^{n} \left(-\frac{\hbar^2}{2} \frac{\delta^2 \Omega_\mu}{\delta\phi_i(x)^2} + \hbar[\phi_i^{tr}(x)(-\nabla^2)^{1/2} + j_i(x)(-\nabla^2)^{-1/2}] \frac{\delta\Omega_\mu}{\delta\phi_i(x)} \right) dx$$

$$= (E_\mu - E_0)\Omega_\mu \tag{K.19}$$

as a consequence of (K.15) and the form of (K.12). Solutions to Eq. (K.19) are readily obtained as polynomial functionals in ϕ^{tr}. For example, the one-photon state solution takes the form

$$\Omega_1 = \int \psi(x) \cdot [\phi^{tr}(x) + (-\nabla^2)^{-1} j(x)] dx \tag{K.20}$$

where $\psi(x)$ is the one-photon wave function, a solenoidal solution to the relativistic wave equation

$$\hbar(-\nabla^2)^{1/2}\psi_i(x) = (E_1 - E_0)\psi_i(x) \tag{K.21}$$

which admits bounded solutions for all finite real positive values of $(E_1 - E_0)$; the two-photon state solution is

$$\Omega_2 = \tfrac{1}{2} \iint [\phi^{tr}(x) + j(x)(-\nabla_x^2)^{-1}] \cdot \psi(x, y) \cdot [\phi^{tr}(y) + (-\nabla_y^2)^{-1} j(y)]$$

$$\times dx\, dy - \tfrac{1}{2}\hbar^2(E_2 - E_0)^{-1} \int \sum_{i=1}^{3} \psi_{ii}(x, x)\, dx \tag{K.22}$$

with the two-photon wave function $\psi(x, y)$ a solenoidal (that is, $\sum_{i=1}^{3} \partial\psi_{ij}(x, y)/\partial x_i = 0 = \sum_{j=1}^{3} \partial\psi_{ij}(x, y)/\partial y_j$) symmetrical (that is, $\psi_{ij}(x, y) = \psi_{ji}(y, x)$) solution to the relativistic wave equation

$$\hbar[(-\nabla_x^2)^{1/2} + (-\nabla_y^2)^{1/2}]\psi_{ij}(x, y) = (E_2 - E_0)\psi_{ij}(x, y) \tag{K.23}$$

---

[1] For a recent and novel derivation of this electromagnetic radiation energy term, see J. Schwinger, *Phys. Rev.* **173**, 1264 (1968).

which admits bounded solutions for all finite real positive values of $(E_2 - E_0)$; solutions to (K.19) describing three or more free photons are given by higher order polynomial functionals in ϕ^{tr}.

This so-called coordinate-diagonal representation for the quantum theory of electromagnetic radiation, with the quantum Hamiltonian prescribed as the functional differential operator (K.12), can also be employed in theories of charged-particle interaction. In particular, the coordinate-diagonal representation has been applied recently to certain basic problems in quantum electrodynamics.[2]

---

[2] G. Rosen, *Phys. Rev.* **172**, 1632 (1968); *Nuovo Cimento* **57**, 870 (1968); *J. Franklin Inst.* **287**, 261 (1969).

Appendix L   Stationary States in Quantum Field Theories

Consider the class of local field theories based on Lagrangian densities of the form

$$\mathscr{L} = \tfrac{1}{2}\dot{\phi} \cdot \dot{\phi} - \tfrac{1}{2}(\nabla\phi) \cdot (\nabla\phi) - u \equiv \tfrac{1}{2} \sum_{i=1}^{n} [(\dot{\phi}_i)^2 - (\nabla\phi_i)^2] - u, \quad (\text{L.1})$$

where ϕ denotes a real n-tuple field amplitude and the self-interaction energy density $u = u(\phi)$ is a real 1-tuple function of ϕ. From the canonically related Hamiltonian for such a theory,

$$H = H[\phi, \pi] = \int (\tfrac{1}{2}\pi \cdot \pi + \tfrac{1}{2}(\nabla\phi) \cdot (\nabla\phi) + u) \, dx, \quad (\text{L.2})$$

we obtain the quantum Hamiltonian

$$\mathbf{H} = H[\phi(x), -i\hbar\delta/\delta\phi(x)]$$

$$= \int \left(-\frac{\hbar^2}{2} \frac{\delta}{\delta\phi(x)} \cdot \frac{\delta}{\delta\phi(x)} + \frac{1}{2} \phi(x) \cdot (-\nabla^2)\phi(x) + u(\phi(x)) \right) dx, \quad (\text{L.3})$$

and for the ϕ-dependent part of a stationary state wave functional $\Psi[\phi; t] = (\exp -iE_\mu t/\hbar)U_\mu[\phi]$, we find the eigenfunctional equation

$$\mathbf{H}U_\mu[\phi] = E_\mu U_\mu[\phi] \quad (\text{L.4})$$

145

with E_μ the constant energy eigenvalue. Physically admissible solutions to Eq. (L.4) are such that $|U_\mu[\phi]|^2$, the relative probability density for locating the state at the field amplitude n-tuple $\phi = \phi(x)$, vanishes for unbounded field magnitudes,

$$\lim_{(\phi \cdot \phi) \to \infty} |U_\mu[\phi]|^2 = 0. \tag{L.5}$$

The $\mu = 0$ vacuum state solution to (L.4) is associated with the energy $E_0 \equiv \min_\mu\{E_\mu\}$; once $U_0[\phi]$ is obtained, the general stationary state eigenfunctional problem reduces to solving the equation

$$\hbar^2 \int \sum_{i=1}^{n} \left(-\frac{1}{2} \frac{\delta^2 \Omega_\mu}{\delta\phi_i(x)^2} - \frac{\delta(\ln U_0[\phi])}{\delta\phi_i(x)} \frac{\delta\Omega_\mu}{\delta\phi_i(x)} \right) dx = (E_\mu - E_0)\Omega_\mu \tag{L.6}$$

for $\Omega_\mu = \Omega_\mu[\phi] \equiv U_\mu[\phi]/U_0[\phi]$ because the left-hand side of (L.6) works out to give

$$\hbar^2 \int \sum_{i=1}^{n} \left(-\tfrac{1}{2} U_0^{-1}[\delta^2 U_\mu/\delta\phi_i(x)^2] + \tfrac{1}{2} U_0^{-2} U_\mu[\delta^2 U_0/\delta\phi_i(x)^2] \right) dx$$

$$= U_0^{-2}(U_0 \mathbf{H} U_\mu - U_\mu \mathbf{H} U_0) \tag{L.7}$$

with $\mathbf{H}$ prescribed by (L.3).

Approximate physical solutions to Eq. (L.4) with the Hamiltonian operator (L.3) have been obtained for certain generic classes of entire analytic $u(\phi)$ by applying the Rayleigh–Ritz procedure for functionalities.[1] In the following, we discuss a class of nonlinear field theories for which exact physical solutions to Eq. (L.4) are obtainable in closed-form. These theories feature a self-interaction energy density that converges rapidly for large field magnitudes to the form representative of a linear field theory of uncoupled field amplitude components, a $u = u(\phi)$ expressible as

$$u = \tfrac{1}{2} \sum_{i=1}^{n} m_i^2 \phi_i^2 + \mathcal{U} \tag{L.8}$$

where $\mathcal{U} = \mathcal{U}(\phi)$ is a continuous singularity-free real 1-tuple function such that

[1] G. Rosen, *Phys. Rev. Let.* **16**, 704 (1966); *Phys. Rev.* **173**, 1680 (1968), and works cited therein.

$$\lim_{|\phi| \to \infty} [\phi^2 \mathscr{U}(\phi)] = 0 \qquad \text{if} \quad n = 1,$$

$$\lim_{(\phi \cdot \phi) \to \infty} [(\phi \cdot \phi)[\ln \phi \cdot \phi] \mathscr{U}(\phi)] = 0 \qquad \text{if} \quad n = 2, \qquad (\text{L}.9)$$

$$\lim_{(\phi \cdot \phi) \to \infty} [(\phi \cdot \phi) \mathscr{U}(\phi)] = 0 \qquad \text{if} \quad n \geqslant 3.$$

For a continuous $\mathscr{U}(\phi)$ that tends to zero for large field magnitudes with the asymptotic behavior prescribed by (L.9), we have existence of the function

$$f(\phi) \equiv \begin{cases} -\displaystyle\int_{-\infty}^{\infty} |\phi - \phi'| \, \mathscr{U}(\phi') \, d\phi' & \text{for} \quad n = 1 \\[2mm] -\dfrac{1}{2\pi} \displaystyle\int [\ln (\phi - \phi') \cdot (\phi - \phi')] \mathscr{U}(\phi') \, d\phi' & \text{for} \quad n = 2 \\[2mm] \dfrac{\Gamma(\frac{1}{2}n - 1)}{2\pi^{n/2}} \displaystyle\int \dfrac{\mathscr{U}(\phi') \, d\phi'}{[(\phi - \phi') \cdot (\phi - \phi')]^{(n/2)-1}} & \text{for} \quad n \geqslant 3 \end{cases}$$

$$(\text{L}.10)$$

that satisfies the n-dimensional Poisson equation

$$\sum_{i=1}^{n} \partial^2 f(\phi)/\partial \phi_i^2 = -2 \mathscr{U}(\phi). \qquad (\text{L}.11)$$

The asymptotic behavior of $f(\phi)$ for large field magnitudes follows from (L.9) and (L.10) as $f(\phi) \propto |\phi|$ if $n = 1$, $f(\phi) \propto [\ln (\phi \cdot \phi)]$ if $n = 2$, and $f(\phi) \propto (\phi \cdot \phi)^{1 - (n/2)}$ if $n \geqslant 3$. In terms of the function (L.10), the $\mu = 0$ vacuum state solution to (L.4) with (L.8) in (L.3) is

$$U_0[\phi]$$

$$= \exp\left(-\frac{1}{2\hbar} \int \sum_{i=1}^{n} \phi_i(x)(-\nabla^2 + m_i^2)^{1/2} \phi_i(x) \, dx - \frac{\varepsilon}{\hbar^2} \int f(\phi(x)) \, dx \right),$$

$$(\text{L}.12)$$

where the limit $\varepsilon \to 0$ is understood to be taken as the final step in a computation involving $U_0[\phi]$, in conjunction with a limit representation of the δ-function $\lim_{\varepsilon \to 0} \delta_{(\varepsilon)}(x) = \delta(x)$ such that[2]

$$\delta_{(\varepsilon)}(0) = \varepsilon^{-1}. \qquad (\text{L}.13)$$

[2] An immediate way to secure this relation is to introduce the wave-number cutoff limit representation

$$\delta_{(\varepsilon)}(x) = \int_{|k| \leqslant K} (\exp ik \cdot x) \, dk/(2\pi)^3$$

with $K \equiv (6 \pi^2/\varepsilon)^{1/3}$.

To verify that (L.12) is an exact solution to (L.4), one simply computes

$$\frac{\delta U_0[\phi]}{\delta \phi_i(x)} = \left\{ -\frac{1}{\hbar}(-\nabla^2 + m_i^2)^{1/2}\phi_i(x) - \frac{\varepsilon}{\hbar^2} \frac{\partial f(\phi)}{\partial \phi_i}\bigg|_{\phi=\phi(x)} \right\} U_0[\phi],$$

(L.14)

$$\frac{\delta^2 U_0[\phi]}{\delta \phi_i(x)^2} = \left\{ -\frac{1}{\hbar}[(-\nabla^2 + m_i^2)^{1/2}\,\delta(x)]_{x=0} - \frac{\varepsilon}{\hbar^2}\delta_{(\varepsilon)}(0)\frac{\partial^2 f(\phi)}{\partial \phi_i^2}\bigg|_{\phi=\phi(x)} \right.$$

$$\left. + \frac{1}{\hbar^2}[(-\nabla^2 + m_i^2)^{1/2}\phi_i(x)]^2 + O(\varepsilon) \right\} U_0[\phi],$$

(L.15)

and makes use of (L.11) and (L.13) to obtain in the limit $\varepsilon \to 0$

$$\int \left\{ \sum_{i=1}^{n} \left(-\frac{\hbar^2}{2}\frac{\delta^2 U_0[\phi]}{\delta \phi_i(x)^2} + \frac{1}{2}[(-\nabla^2 + m_i^2)^{1/2}\phi_i(x)]^2 U_0[\phi] \right) \right.$$

$$\left. + \mathscr{U}(\phi(x))U_0[\phi] \right\} dx = E_0\, U_0[\phi], \quad \text{(L.16)}$$

where the (unobservable) vacuum state energy appears as the infinite constant

$$E_0 = (\hbar/2)\sum_{i=1}^{n}[(-\nabla^2 + m_i^2)^{1/2}\delta(x)]_{x=0}\int dx. \quad \text{(L.17)}$$

By substituting (L.12) into Eq. (L.6), we find

$$\int \sum_{i=1}^{n} \left(-\frac{\hbar^2}{2}\frac{\delta^2 \Omega_\mu}{\delta \phi_i(x)^2} + \left[\hbar\phi_i(x)(-\nabla^2 + m_i^2)^{1/2} + \varepsilon\frac{\partial f}{\partial \phi_i}\bigg|_{\phi=\phi(x)} \right]\frac{\delta \Omega_\mu}{\delta \phi_i(x)} \right) dx$$

$$= (E_\mu - E_0)\Omega_\mu, \quad \text{(L.18)}$$

an equation which facilitates determination of excited state eigen-functional solutions to (L.4), $U_\mu[\phi] = U_0[\phi]\Omega_\mu[\phi]$ with $E_\mu > E_0$. Solutions to (L.18) are continuous in ε about $\varepsilon = 0$, and therefore the limit $\varepsilon \to 0$ obtains validity in the equation itself. Hence, the field theory is effectively linear with two-particle and multiparticle state solutions exhibiting no interaction between the quanta.

Index

A

Action
  functional, 7, 11, 55, 59
  principle, 8, 11, 56, 59
Algebra, *see* Lie algebra
Anticommutation relations, 84

C

Canonical equations, *see* Hamilton
  canonical equations
Canonical transformations, 12, 16–20,
  46–50, 63–66, 88–91
Canonical variables, 9, 11, 30, 41, 59
Characteristic function, 130
Chronological ordering, 40–41, 79–80,
  132–133
Class, 7, 96
Constants of motion, 13, 42, 62, 85,
  87
Contact transformations, *see* Canoni-
  cal tranformations

D

Dilatation transformations, 56, 134–
  136, 138–139

D

Dirac, 94
  bracket, 42, 85
  correspondence, 43–45, 86–87
  electron, 40–41, 84, 118–122
Dynamical law, 5, 52
Dynamical principle of superposition,
  22, 68
Dynamics, 5, 22, 52

E

Einstein, 95
  field theories, 134
Electrodynamics, quantum, 93–94,
  144
Electromagnetic radiation, quantum
  theory of, 140–144
Energy
  canonical, 6, 54, 57
  density, self-interaction, 53
  eigenstates of, *see* Stationary states
  eigenvalues, 38–39, 81–82
  multiplets, degenerate, 39
  observable, 83

Energy (*cont.*)
  potential, 5
  vacuum, 81
Enveloping algebra, 13
Euler–Lagrange equations, 5, 30
  for fields, 52, 74
  Newtonian form of, 5
Expectation values, 38, 80, 127

F

Feynman operators, 29–30, 74–75,
  123–126
Feynman postulate, 24, 30, 70
Field
  amplitude, 52
  canonical variables for, 57
  linear, 53
  local, 53
  momentum density, 57
  nonlinear, 53, 56
  semilinear, 54
  solutions for, 134–139
  velocity, 52
Functional, 96
  analytic, 99
  continuous, 96
  derivative, 97–99
  extremum of, 98
  integral, 24–27, 70–71
  integration by parts, 30, 116–117
  Laplacian, 99
  linear, 97

G

Generalized coordinate *n*-tuple, 4
Generalized momentum *n*-tuple, 8
Generalized velocity *n*-tuple, 4
Generators, 19, 37, 48, 64–65, 101
Green's function, causal, 75, 126

H

Haar measure, 25–27, 69, 71, 110–115
Hamilton canonical equations, 9

Hamilton canonical field equations,
  57
Hamiltonian, 9, 57
Heisenberg commutation relations, 126
Heisenberg picture, 41, 85
Hypercharacteristic function, 128–129

I

Interaction wave function, 131
Interaction wave functional, 79
Iteration solution, 29, 37, 74, 79, 131–
  133

K

Kepler Hamiltonian, 46

L

Lagrangian, 1–6, 52, 54
  density, 53
Legendre transform, 9, 10
Lie algebras, 13–16, 42–46, 61–62,
  87–91
  conjugate, 14–15
Lie groups, 19, 101–109
Lie product, 13
Lie quadratic identities, 19, 102

N

n-tuple, 4
Normalization, 25, 27, 36, 70

O

Observables, 12, 41, 46, 60, 85

P

Pauli neutrino, 40–41, 84
Phase space, 18
Planck's constant, 24
Poisson bracket, 12–13, 17, 60
Probability density, 22, 68
Propagation kernel, 23–24, 69–70

Q

Quantum Hamiltonian, 37, 40, 79
Quasiclassical approximation, 33–34, 77

R

Rayleigh–Ritz procedure for functionalities, 146
Relativistic sum-over-histories, 118–122
Renormalization, 93

S

S-matrix, 80
Schrödinger equation, 37–38, 78–79
Semidirect sum, 14
Semigroup composition law, 23, 69
Space, 7, 97
State, 4, 9, 22, 52, 57, 68
   ground, 39
Stationary states, 38, 80–83, 142–143, 145–148

Statistical mechanics, 127–130
Structure constants, 19, 48, 102
Symmetry algebra, 21, 50, 66
Symmetry group, 21, 50, 66, 92
Symplectic group, 14

T

Topological group, 26, 71

U

U-matrix, 80

V

Volterra expansion, 33, 99

W

Wave function, 22, 34
Wave functional, 68
Wiener correlation function, 117
Wiener functional integral, 25, 117

Mathematics in Science and Engineering

A Series of Monographs and Textbooks

Edited by RICHARD BELLMAN, *University of Southern California*

1. T. Y. Thomas. Concepts from Tensor Analysis and Differential Geometry. Second Edition. 1965

2. T. Y. Thomas. Plastic Flow and Fracture in Solids. 1961

3. R. Aris. The Optimal Design of Chemical Reactors: A Study in Dynamic Programming. 1961

4. J. LaSalle and S. Lefschetz. Stability by by Liapunov's Direct Method with Applications. 1961

5. G. Leitmann (ed.). Optimization Techniques: With Applications to Aerospace Systems. 1962

6. R. Bellman and K. L. Cooke. Differential-Difference Equations. 1963

7. F. A. Haight. Mathematical Theories of Traffic Flow. 1963

8. F. V. Atkinson. Discrete and Continuous Boundary Problems. 1964

9. A. Jeffrey and T. Taniuti. Non-Linear Wave Propagation: With Applications to Physics and Magnetohydrodynamics. 1964

10. J. T. Tou. Optimum Design of Digital Control Systems. 1963.

11. H. Flanders. Differential Forms: With Applications to the Physical Sciences. 1963

12. S. M. Roberts. Dynamic Programming in Chemical Engineering and Process Control. 1964

13. S. Lefschetz. Stability of Nonlinear Control Systems. 1965

14. D. N. Chorafas. Systems and Simulation. 1965

15. A. A. Pervozvanskii. Random Processes in Nonlinear Control Systems. 1965

16. M. C. Pease, III. Methods of Matrix Algebra. 1965

17. V. E. Benes. Mathematical Theory of Connecting Networks and Telephone Traffic. 1965

18. W. F. Ames. Nonlinear Partial Differential Equations in Engineering. 1965

19. J. Aczel. Lectures on Functional Equations and Their Applications. 1966

20. R. E. Murphy. Adaptive Processes in Economic Systems. 1965

21. S. E. Dreyfus. Dynamic Programming and the Calculus of Variations. 1965

22. A. A. Fel'dbaum. Optimal Control Systems. 1965

23. A. Halanay. Differential Equations: Stability, Oscillations, Time Lags. 1966

24. M. N. Oguztoreli. Time-Lag Control Systems. 1966

25. D. Sworder. Optimal Adaptive Control Systems. 1966

26. M. Ash. Optimal Shutdown Control of Nuclear Reactors. 1966

27. D. N. Chorafas. Control System Functions and Programming Approaches (In Two Volumes). 1966

28. N. P. Erugin. Linear Systems of Ordinary Differential Equations. 1966

29. S. Marcus. Algebraic Linguistics; Analytical Models. 1967

30. A. M. Liapunov. Stability of Motion. 1966

31. G. Leitmann (ed.). Topics in Optimization. 1967

32. M. Aoki. Optimization of Stochastic Systems. 1967

33. H. J. Kushner. Stochastic Stability and control. 1967

34. M. Urabe. Nonlinear Autonomous Oscillations. 1967

35. F. Calogero. Variable Phase Approach to Potential Scattering. 1967

36. A. Kaufmann. Graphs, Dynamic Programming, and Finite Games. 1967

37. A. Kaufmann and R. Cruon. Dynamic Programming: Sequential Scientific Management. 1967

38. J. H. Ahlberg, E. N. Nilson, and J. L. Walsh. The Theory of Splines and Their Applications. 1967

39. Y. Sawaragi, Y. Sunahara, and T. Nakamizo. Statistical Decision Theory in Adaptive Control Systems. 1967

40. R. Bellman. Introduction to the Mathematical Theory of Control Processes Volume I. 1967 (Volumes II and III in preparation)

41. E. S. Lee. Quasilinearization and Invariant Imbedding. 1968

42. W. Ames. Nonlinear Ordinary Differential Equations in Transport Processes. 1968

43. W. Miller, Jr. Lie Theory and Special Functions. 1968

44. P. B. Bailey, L. F. Shampine, and P. E. Waltman. Nonlinear Two Point Boundary Value Problems. 1968.

45. Iu. P. Petrov. Variational Methods in Optimum Control Theory. 1968

46. O. A. Ladyzhenskaya and N. N. Ural'tseva. Linear and Quasilinear Elliptic Equations. 1968

47. A. Kaufmann and R. Faure. Introduction to Operations Research. 1968

48. C. A. Swanson. Comparison and Oscillation Theory of Linear Differential Equations. 1968

49. R. Hermann. Differential Geometry and the Calculus of Variations. 1968

50. N. K. Jaiswal. Priority Queues. 1968

51. H. Nikaido. Convex Structures and Economic Theory. 1968

52. K. S. Fu. Sequential Methods in Pattern Recognition and Machine Learning. 1968

53. Y. L. Luke. The Special Functions and Their Approximations (In Two Volumes). 1969

54. R. P. Gilbert. Function Theoretic Methods in Partial Differential Equations. 1969

55. V. Lakshmikantham and S. Leela. Differential and Integral Inequalities (In Two Volumes). 1969

56. S. H. Hermes and J. P. LaSalle. Functional Analysis and Time Optimal Control. 1969.

57. M. Iri. Network Flow, Transportation, and Scheduling: Theory and Algorithms. 1969

58. A. Blaquiere, F. Gerard, and G. Leitmann. Quantitative and Qualitative Games. 1969

59. P. L. Falb and J. L. de Jong. Successive Approximation Methods in Control and Oscillation Theory. 1969

60. G. Rosen. Formulations of Classical and Quantum Dynamical Theory. 1969

61. R. Bellman. Methods of Nonlinear Analysis, Volume I. 1970

62. R. Bellman, K. L. Cooke, and J. A. Lockett. Algorithms, Graphs, and Computers. 1970

In preparation

A. H. Jazwinski. Stochastic Processes and Filtering Theory

S. R. McReynolds and P. Dyer. The Computation and Theory of Optimal Control

J. M. Mendel and K. S. Fu. Adaptive, Learning, and Pattern Recognition Systems: Theory and Applications

E. J. Beltrami. Methods of Nonlinear Analysis and Optimization

H. H. Happ. The Theory of Network Diakoptics

M. Mesarovic, D. Macko, and Y. Takahara. Theory of Hierarchical Multilevel Systems